Hymns for the Fallen

The publisher gratefully acknowledges the generous support of the Constance and William Withey Endowment Fund for History and Music of the UC Press Foundation.

Hymns for the Fallen

Combat Movie Music and Sound after Vietnam

TODD DECKER

University of California Press

University of California Press, one of the most distinguished university presses in the United States, enriches lives around the world by advancing scholarship in the humanities, social sciences, and natural sciences. Its activities are supported by the UC Press Foundation and by philanthropic contributions from individuals and institutions. For more information, visit www.ucpress.edu.

University of California Press
Oakland, California

Library of Congress Cataloging-in-Publication Data
Names: Decker, Todd R., author.
Title: Hymns for the fallen : combat movie music and sound after Vietnam / Todd Decker.
Description: Oakland, California : University of California Press, [2017] | Includes bibliographical references and index.
Identifiers: LCCN 2016034599 (print) | LCCN 2016038416 (ebook) | ISBN 9780520282322 (cloth : alk. paper) | ISBN 9780520282339 (pbk. : alk. paper) | ISBN 9780520966543 (ebook)
Subjects: LCSH: Film soundtracks—History and criticism. | Motion picture music—History and criticism. | War films—History and criticism.
Classification: LCC ML2075 .D43 2017 (print) | LCC ML2075 (ebook) | DDC 781.5/420973--dc23
LC record available at https://lccn.loc.gov/2016034599

Manufactured in the United States of America

26 25 24 23 22 21 20 19 18 17
10 9 8 7 6 5 4 3 2 1

For David.
He always makes the popcorn.

Contents

War has its own conventions.

—J. GLENN GRAY, *The Warriors:*
Reflections on Men in Battle (1958)

Very little is seen in war, anyway. Wars are fought by ear.

—HARRY BROWN, *A Walk in the Sun* (1944)

You know, the trouble with war is that there isn't any background
music.

—PHILIP CAPUTO, *A Rumor of War* (1977)

Sergeant Bercaw covered his retreating soldiers by rising to his
knees and firing magazine after magazine on full automatic ("full
rock 'n' roll" in soldier parlance).

—JOHN C. MCMANUS, *Grunts* (2010)

We sat together for a few minutes, listening to the cadence of
empire. Cargo planes rumbled through the winter sky while helos
sliced at it. Soldiers made jokes about small dicks and big dicks.
With the sun falling in the west, someone in the gulch trilled the
Zulu chant from the beginning of *The Lion King*.
    I smiled in spite of myself.

—MATT GALLAGHER, *Youngblood* (2016)

Introduction

One night in 1979, Vietnam veteran Jan Scruggs went to see director Michael Cimino's *The Deer Hunter*. The film stirred Scruggs's deep memories of the war. After a flashback at three o'clock in the morning to his own combat experience and the men who died beside him, Scruggs realized, "No one remembers their names." He resolved the next morning "to build a memorial to all the guys who served in Vietnam. It'll have the name of everyone killed."[1] And so, one veteran's experience of a war movie inspired the making of the Vietnam Veterans Memorial on the National Mall in Washington, DC. Designed by Maya Lin, the Wall—as it quickly came to be known—was dedicated in 1982 and soon thereafter began appearing in an ongoing cycle of Hollywood films about the Vietnam War—a cycle initiated, in part, by *The Deer Hunter*.

The 1987 film *Hamburger Hill* opens at the Wall after an initial, entirely sonic evocation of the Vietnam War. Over white titles on a black background, we hear a radio call for help: a unit taking fire requests assistance. They are promptly answered in the affirmative, after which the sound of a helicopter rotor—perhaps heading toward the embattled platoon—enters the mix. Informative titles set the date (May 1969) and location (Hill 937 in the Ashau Valley) as music by the minimalist art music composer Philip Glass fades in: a bubbling, rhythmic music, urgent and dark, featuring repeated minor-mode figures in the low strings with bursts of percussion and brass. All three elements of the soundtrack—dialogue, sound effects, and music—are activated in *Hamburger Hill* before a single image appears. Sound alone puts the viewer onto the battlefield, in a soundscape where music has a place beside radio chatter and noisy war machines.

Images begin to flash in alternation with title cards as Glass's music alone plays on: the US Capitol seen from near the Wall; a black wreath against the Wall's black surface; a view down the Wall's sloping pathway, uncharacteristically empty of visitors on a wintry, windswept day. The fourth image sets *Hamburger Hill* in motion. The camera tracks along the surface of the Wall; the names of the dead slide by at a speed no walker could hope to match. On successive cuts back to the Wall, the names appear larger. The combination of shot choice and music at *Hamburger Hill*'s opening effectively transforms the then still-new memorial into a cinematic experience. We are both at the Wall—the location is undeniably real—and experiencing the Wall as film. (For some among *Hamburger Hill*'s original audiences, their first experience of the Wall was at the movies.) As the credits come to an end, sound effects, radio voices, and gunfire reenter the mix, eventually displacing the music, which falls silent. The tracking shot along the Wall slowly cross-fades into a similarly paced tracking shot following American soldiers moving through the high grass of Vietnam. We move through a cinematic representation of the Wall into *Hamburger Hill*'s representation of the war. As one reviewer noted, "More than any of the films to come out about Vietnam, *Hamburger Hill* wants to be a memorial to our experience there—a cinematic headstone."[2]

Hamburger Hill was written by Vietnam veteran Jim Carabatsos, whose original script opened with a dramatic scene at the Wall not included in the film. A young "AMERICAN FAMILY" walks toward the monument: a father and mother in their mid-thirties with two small children. In Carabatsos's words, "The FATHER has his back to us (we never see him). He stops in front of the Memorial. . . . We can *feel* the emotion coming from the man."[3] On the soundtrack, sounds of the present (the children say, "Daddy's crying") and the past ("a staccato, STATIC-FILLED RADIO language") overlap. As the children approach him, the father "[senses] them next to him and the RADIO VOICES STOP." Sound, integral to the screenwriter's conception, gives the viewer privileged access to the inner life of this man, who is, by implication, a Vietnam veteran. Carabatsos's script also closes at the Wall: "The Father stands straighter . . . prouder . . . and helps his son plant a small American flag."

Hamburger Hill as released does not end at the Wall but instead concludes with a long, quiet battlefield coda. The soundtrack goes almost completely silent once the objective of the battle—the strategically meaningless Hill 937—is taken. For four minutes the viewer sits in near silence, watching sustained close-ups of the three American soldiers who survive, one of whom slowly sheds a single tear (figure 1). No music guides our reflection on the film; no rounded melody hints at when this endless shot might conclude. Finally, helo sounds and radio chatter fade in over informative title

FIGURE 1. A soldier's tear: surveying the wasted landscape of *Hamburger Hill* and the "insane" American war in Vietnam

cards that sum up the battle and note, "The war for hills and trails continued, the places and names forgotten, except by those who were there"—some of whom, no doubt, were in *Hamburger Hill*'s original audiences. A second text follows: a 1970 poem by Major Michael Davis O'Donnell scrolls upward as Glass's disturbing music, heard at the top of the film, returns. O'Donnell's poem asks the reader to "embrace those gentle heroes you left behind" once "men decide and feel safe to call the war insane." Glass's music plays on, without abating in intensity, to the close of the credit roll. Critics heard this music as "stringent and grim," "teaming and sobering": the memory of Vietnam in 1987 was fresh and still troubling.[4]

The framing sequences with the young family in the draft script for *Hamburger Hill* eerily anticipate the opening and closing scenes of director Steven Spielberg's World War II film *Saving Private Ryan* (1998)—a celebrated movie about a war the cultural memory of which stands in tremendous contrast with that of Vietnam. While *Hamburger Hill* opens at a memorial erected in the symbolic space of the National Mall, *Saving Private Ryan* begins on sacred blood-soaked ground: the American military cemetery at Omaha Beach in Normandy, where the names of the D-Day dead are listed one by one on tombstones. An aged veteran—by inference a veteran of the war—walks well ahead of his wife, grown son and daughter-in-law, and several grandchildren. The old man moves alone among the graves, finds the one he's looking for, and falls to his knees. The only line of dialogue in the sequence is his son's cry, "Dad." Otherwise, composer John

Williams's orchestral score carries the soundtrack almost entirely, alternating between restrained and sober brass fanfare-like figures and achingly sad string lines. Occasionally clarinets—a Williams favorite for moments of comforting—come to the old man's aid. The movement back in time to the film's D-Day landing battle sequence is abrupt, prepared only on the soundtrack, which shifts from a dissonant, dynamically growing tone cluster in the score to the crashing waves on Omaha Beach. Score yields to effects on a hard cut to the past.

At the close, *Saving Private Ryan* returns to the old man at the cemetery, revealed to be Ryan himself. Again, music mostly carries the soundtrack, except for a rather mawkish spoken exchange between Ryan and his wife. The narrative goes to black and silence on a drawn-out final cadence in the score, after which the end titles roll to the sound of a musical benediction by Williams, who composed a six-minute piece for orchestra and chorus unrelated thematically to the rest of the score and titled "Hymn to the Fallen." On the soundtrack CD, produced by Williams, "Hymn to the Fallen" is included two times: as the first and last track, with the film's dramatic score sandwiched between. For the home listener, "Hymn to the Fallen" serves as prelude and postlude to reflection on Spielberg's film by way of Williams's score. And it works in the concert hall as well. I heard "Hymn to the Fallen" on an all-Williams concert given by the St. Louis Symphony Orchestra in 2013, and the audience's reaction to it was markedly different from its response to the other pieces on the program. "Hymn to the Fallen" was received as more than just movie music: it carried a larger meaning. One could applaud not just Williams and the orchestra and chorus but also veterans, soldiers, and the sacred idea of sacrifice for the nation.

In his 2009 book *Monument Wars*, Kirk Savage describes the public monuments on the National Mall in Washington, DC, as speaking "to a deep need for attachment that can be met only in a real place, where the imagined community actually materializes and the existence of the nation is confirmed in a simple but powerful way."[5] Another place where the imagined community of the nation materializes is in the movie theater, where war films—especially the thematically serious war films made in the decades following the conclusion of the Vietnam War, grouped here under the sub-genre rubric *prestige combat films*—have served a monumental function as sites of shared access to greater truths about the nation, specifically through the figures of the soldier and the veteran. This book describes in detail how music and sound function as constituent parts of the prestige combat film's larger work of memorialization in the cultural realm of commercial cinema. As Rikke Schubart and Anne Gjelsvik note, historians must deal with "the complexity of history, war, heroism, patriotism, memory, and the process of their representation."[6] *Hymns for the Fallen* traces an expressive sonic continuity in this "process of representation" for serious war films.

The three elements of the soundtrack—dialogue, sound effects, music— are treated in detail in the chapters that follow, although music proves to be of particular interest. Indeed, the prestige combat subgenre is thoroughly musical, much more so than the war films made by Hollywood before the Vietnam War. Parts 2, 3, and 4 of this book each take up an element of the soundtrack in turn. While dividing the analysis into these larger domains, the overall soundtrack mix remains a fundamental frame of reference. Each of the three elements only ever functions in the presence or absence of the other two, and I try throughout to account for this dynamic and its effect on sonic and musical meaning: listening selectively—as screenwriters, sound designers, and composers do—while also keeping the whole mix in mind—as sound mixers and directors do.

Part 2 considers dialogue, focusing in turn on how the soldiers in these films talk (chapter 3), on soldiers' singing, listening, and talking about popular music (chapter 4), and on various sorts of disembodied voices (chapter 5). Part 3 considers sound effects: chapter 6 surveys the meaning-making sound of specific weapons and the mixing of battle scenes; chapter 7 compares sonic realizations of the helicopter. Part 4 deals with music. Here the differences between pre- and post-Vietnam war films are profound. Indeed, the American experience in Vietnam—the national ordeal of losing a war— effectively forced filmmakers and composers to create new musical conventions for the war film. Music in combat films about the Vietnam War made during the long cultural wake of that war—from the late 1970s to the end of the 1980s—demanded a sudden shift in genre conventions. Significantly, these innovative war movie music conventions crafted for narratives about Vietnam were then applied to other wars, from World War II to the Gulf War to the so-called Global War on Terror. Chapter 8 considers several kinds of unmetered musics that stand in stark contrast to the pre-Vietnam war film score paradigm of the march. Chapter 9 looks at triple-meter or waltz-time scores (heard in World War II films) and beat-driven electronic scores (used in films depicting the United States in the Middle East). Chapter 10 explores the single most important musical innovation in war movie music: the elegiac register, a kind of movie music originating in Oliver Stone's *Platoon* (1986) and resounding into the present. The book concludes with a consideration of how music for the end titles has been used to close out nearly all prestige combat films in a reflective, fundamentally memorializing mode.

TABLE 1. The Prestige Combat Film Genre

1978	*The Boys in Company C*
	Go Tell the Spartans
	The Deer Hunter
1979	*Apocalypse Now*
1986	*Platoon*
1987	*Hamburger Hill*
	Full Metal Jacket
1989	*84 Charlie MoPic*
	Casualties of War
	Born on the Fourth of July
1996	*Courage Under Fire*
1998	*Saving Private Ryan*
	The Thin Red Line
1999	*Three Kings*
2001	*Black Hawk Down*
	Band of Brothers
2002	*We Were Soldiers*
	Windtalkers
2005	*Jarhead*
2006	*Flags of Our Fathers*
	Letters from Iwo Jima
	United 93
2007	*In the Valley of Elah*
	Redacted
	The Kingdom
2008	*Miracle at St. Anna*
	The Hurt Locker
	Generation Kill
2010	*Green Zone*
	The Pacific
2012	*Red Tails*
	Act of Valor
	Zero Dark Thirty
2013	*Lone Survivor*
2014	*American Sniper*

Before these more focused discussions, part 1 contains two subgenre-framing chapters, each offering a broad overview of the thirty-five films grouped together as prestige combat films in this study. (Table 1 provides a chronological list of the films.) Chapter 1 sketches out important shared characteristics of prestige combat films outside of sonic matters. This topical and chronological overview grounds subsequent analyses of the soundtrack in industry, genre, visual style, and reception history and speaks in more detail to the memorial function of the subgenre. Topics discussed include the importance of the Vietnam War as a national trauma activating a necessary change in the Hollywood war film, the explicitly articulated serious intent of prestige combat film makers, the importance of authenticity (variously defined), and the ambiguous reception of these films by different audiences, especially young men. The chapter concludes with a brief profile of the four partially overlapping prestige combat film production cycles. Each cycle related differently to the changing figures of the soldier and the veteran. Chapter 2 presents a large-scale comparison of the films' sonic and musical content, offering a bird's-eye view, as it were, that reveals the musical nature of the subgenre, draws aesthetic connections between individual films, and introduces the book's approach to film music and sound more generally. Discussions of specific topics in later chapters are pointed to parenthetically throughout.

Across the book, as in the paired analysis of *Hamburger Hill* and *Saving Private Ryan* opening this introduction, I draw on archival evidence (such as draft screenplays), media and scholarly discourse, ancillary texts (such as score albums), and close readings of image and sound tracks. A comparative approach predominates. Famous, widely discussed films—such as *Apocalypse Now* (1979), *Platoon*, *Saving Private Ryan*, *The Thin Red Line* (1998), *Black Hawk Down* (2001), *The Hurt Locker* (2008), and *Zero Dark Thirty* (2012)—are analyzed from different angles in multiple chapters as the book's focus shifts from dialogue to effects to music. Lesser-known movies—such as *Go Tell the Spartans* (1978), *Courage Under Fire* (1996), *United 93* (2006), and *In the Valley of Elah* (2007)—as well as cinematically produced cable television series—*Band of Brothers* (2001), *The Pacific* (2010), *Generation Kill* (2008)—are treated beside the signal war films of the last four decades. Readers interested in following a particular title through the book are directed to the index, where page numbers in bold connote sustained discussions of a given film.

PART I

The Prestige Combat Film

1. Movies and Memorials

At the most basic level of shared content, prestige combat films—hereafter PCFs—tell stories of US soldiers fighting abroad in actual historical conflicts. (*United 93* [2006] and *Letters from Iwo Jima* [2006] are the exceptions.) Feature films about the American Civil War, which lack a foreign other, and fantasies of American forces at war with imagined enemies (for example the alien invaders of *Independence Day* [1996]) are excluded. Likewise excluded are movies that depict the US military in a fantastical context, such as *Rambo: First Blood, Part Two* (1985), which returns to Vietnam to rescue POWs and, in the words of John Rambo, "win this time," and *Top Gun* (1986), which elides entirely the dire seriousness that would have attended a dogfight between American F-14s and Communist MiGs in the 1980s and instead celebrates winning, as Christian Appy aptly notes, "a fictional battle in an unknown place against a nameless enemy with no significant cause at stake."[1] PCF narratives engage seriously with historical fact—in only a few cases by way of highly stylized storytelling—and insert the viewer, assumed to be an adult, into a complex context. As the director Oliver Stone said, hopefully, of *Platoon* (1986) two years after its release: "It became an antidote to *Top Gun* and *Rambo*."[2]

This complex context, however, is limited in scope. Nearly every PCF represents the battlefield from the point of view of the individual soldier, frequently from the lowest rank: the grunt. Central characters in these films seldom rise above lieutenant (with leading roles in *Saving Private Ryan* [1998], *Band of Brothers* [2001], *We Were Soldiers* [2002], and *Green Zone* [2010] notable exceptions). The PCF is generally not about officers, and never about famous figures of military history—as, for example, were many war films made during the 1960s. Jay Winter has located this larger shift in war films post-1970 as one from "studies of conflict to studies of combatants."[3] To borrow the words of the military historian John C. McManus, the PCF typically strives to capture "the very essence of the infantryman's decidedly personal war."[4] As Stone said rather precisely of *Platoon*, "I did a white Infantry boy's view of the war."[5]

The Vietnam Veterans Memorial marked a radical departure from earlier war memorials in the nation's capital. Kirk Savage characterizes the Wall, which is sunk below grade, as "almost literally [turning] the neoclassical memorial landscape [of the Lincoln, Washington, and other memorials] upside down."[6] Many PCFs about Vietnam did the same, redirecting the heroic narratives of the combat film, as forged during and after World War II, toward the telling of a war story that, in the case of Vietnam, ends in failure and defeat, a deeply ambiguous outcome for a nation as accustomed to victory as the United States. As John Hellmann has noted, Vietnam marks "the disruption of the American story."[7] Katherine Kinney adds, "Vietnam is the traumatic site which violates all images and assumptions of American identity."[8] Or as Michael Herr put it in his 1977 Vietnam memoir, *Dispatches*—zeroing in on the sense of national shame with not a trace of sentimentality—"There's nothing so embarrassing as when things go wrong in a war."[9] Disruption, trauma, and shame are all manifest in most PCFs made after Vietnam—regardless of the war they depict. As David Kieran has argued, "The evolving and contested memory of the American War in Vietnam has shaped Americans' commemoration of other events in ways that inform their understanding of themselves, the nation, and the global interests and obligations of the United States."[10] The Hollywood war film was also shaped by the events and outcomes of the Vietnam War: the PCF, especially in its sonic dimensions, offers a rich space to explore how the experience of Vietnam has resonated across American memory.

And the memory these films build is explicitly national. The media scholars Karina Aveyard and Albert Moran have noted, "Watching a film is also about the people with whom the experience is shared, as well as the moment in time and the place in which it occurs."[11] PCFs are parochial and often occasional: their assumed audience is American (with the exceptions of *Full Metal Jacket* [1987] and *The Thin Red Line* [1998], and perhaps British director Sam Mendes's *Jarhead* [2005]). Hollywood's commercially oriented address to a global audience is largely set aside in the PCF subgenre.

War memorials and PCFs alike recognize the sacrifices soldiers make for the nation. The experience of viewing these films—the time spent watching, especially when done collectively in a movie theater—becomes a constituent part of the viewer's specifically American identity, somewhat like a journey to the Mall in Washington, DC. A majority of PCFs make room for—spend valuable screen time on—explicitly memorializing sequences. Some, like *Hamburger Hill* (1987), *Saving Private Ryan,* and the Vietnam film *We Were Soldiers,* visit real memorials. *We Were Soldiers,* based closely on the battle of Ia Drang, ends at the Wall. Lieutenant Colonel Harold G. Moore—the officer in command at Ia Drang, played in the film by Mel Gibson—stands before the panel where the names of his soldiers killed in the battle are listed. Their names, familiar by now to the viewer as characters in the film, are shown and a title card pinpoints the location of the American dead at Ia Drang on the Wall, implicitly inviting the audience to go and stand in Moore's—and Gibson's—place. If they cannot, watching *We Were Soldiers* serves as a surrogate act of remembrance.

Other combat films memorialize on-screen the names of fallen soldiers who have yet to be remembered in stone in the nation's capital. The 2001 film *Black Hawk Down*—like *We Were Soldiers,* made before but released after the terrorist attacks of September 11, 2001—lists the names of the Army Rangers and members of Delta Force who died on a single day in 1993 in Mogadishu, Somalia. *Act of Valor,* a 2012 film starring actual Navy SEALs, closes with a dedication "to the following warriors of Naval Special Warfare who have made the ultimate sacrifice since 9/11." Sixty names scroll upward while restrained, quiet music plays and an actual Navy SEAL—one of the leading actors in the film, a real soldier who plays a fictional soldier—exits into the sunset. All of the above films, like *Hamburger Hill* though with different motivations, aspire to being a kind of "cinematic headstone."

Some war films go beyond listing names and add images of the fallen and those who survived. Clint Eastwood's *Flags of Our Fathers* (2006) tells the story of the six flag raisers in the iconic 1945 photo of Marines atop Mount Suribachi on the island of Iwo Jima. During the final credits, the names of the actors who played these men are listed beside photos of the actual men. The HBO limited series *Band of Brothers,* which recounts the combat service of a celebrated unit of paratroopers in Europe during World War II, includes actual veterans of the unit in documentary-style interviews at the start of almost every episode. With even greater impact, *Lone Survivor* (2013), an account of Operation Red Wings in the mountains of Afghanistan, closes with images of the nineteen Navy SEALs and Special Operations aviators who died on a single day in 2005. The images are personal, and in the context of a feature film, uncomfortably intimate.

Films incorporating images of actual soldiers and veterans intensify a common trope in Hollywood combat films reaching back to the beginnings of the genre: films such as *Battleground* (1949) and *To Hell and Back* (1955) enhance their closing credits with a visual roll call of the cast, one final glimpse of each man in the film's story. Almost all of the combat films about Vietnam made in the 1980s incorporate this old war movie device, as do several later PCFs about other wars.[12] The visual roll call that ends *Platoon* left many Vietnam veterans in tears—a common human-interest story in local newspapers during the film's theatrical release. Other strategies for initiating reflection include didactic titles at the start or close, as well as stretches of reflective music, such as John Williams's "Hymn to the Fallen" in *Saving Private Ryan*.

Almost all of the above strategies for honoring individual fighting men stop the action narrative's forward motion—or put off the film's end—and force the audience to reflect, thereby opening a cinematic space where soldiers and veterans as embodiments of the nation are shown to be worthy of a memorializing moment's pause.

The action-adventure genre has dominated Hollywood's business model since the mid-1970s, around the time the PCF emerged. Indeed, the PCF—with its de rigueur inclusion of violent, frequently spectacular combat action—is without a doubt an action-adventure subgenre. But while standard commercial action films might set ever-higher box-office records, they typically earn low marks, if not utter contempt, from critics and seldom win anything but technical awards at the Oscars. PCFs, by contrast, manage to be both action films and critical successes judged worthy of major awards, recognition that buttresses the subgenre's claim to prestige. This book considers three winners of and seven nominees for the Academy Award for Best Picture, and five winners and five nominees for the Academy Award for Best Director. Four Oscar-nominated original scores are represented as well. Interestingly for this study, PCFs also often win in the sound categories. Six signature PCFs, each definitive for the subgenre in its period, won Best Sound Mixing Oscars: *The Deer Hunter* (1978), *Apocalypse Now* (1979), *Platoon, Saving Private Ryan, Black Hawk Down, The Hurt Locker* (2008). This startling pattern suggests the centrality of sound in post-Vietnam combat films. (Before 1977, only two war films won this award: *Patton* [1970] and *Twelve O'Clock High* [1949]). Best Sound Design Oscars—a more occasional award for the early decades of the subgenre—were won by *Saving Private Ryan, Letters from Iwo Jima, The Hurt Locker, Zero Dark Thirty* (2012), and *American Sniper* (2014).

PCFs are typically special projects initiated by a director or a producer—less often a writer or actor—working anywhere in the commercial feature industry: inside or outside the studios, at any level of budget, and in the twenty-first century expanding into premium cable television. The cachet of the creative artist behind a given film necessarily determines the scale of the project. This study finds extravagant and modest films talking to each other aesthetically in startling ways.

Most PCF makers are driven by a desire to represent American soldiers at war in a serious manner that contributes to the larger, ever-changing national conversation around soldiers and veterans. Indeed, evidence for such an effort on the part of producers and directors qualifies as a defining aspect of the subgenre, a crucial element in the process of how these films come to be made and their claims to importance. Preproduction pitches, press packs, publicity, and media discourses consistently present PCFs as more than mere movies. The *St. Louis Post-Dispatch* titled its review for *Courage Under Fire* (1996) "An Action Flick for Thinking People," aptly characterizing the intent behind PCFs on the whole.[13] Hong Kong action director John Woo was attracted to *Windtalkers* (2002) by the chance to make, as described in the film's press pack, "a character-driven, emotional action drama" that was, in Woo's words, "so emotional, a celebration of the human spirit . . . something different from a generic action film."[14] So, too, most all PCFs, even those offering a kind of negative image of the human spirit (such as *Full Metal Jacket*).

The PCF often springs from a sense of moral urgency, typically in response to veterans and their families. Jim Carabatsos's script for *Hamburger Hill* bounced around Hollywood for years before producer Marcia Nasatir took it up, in part because her son had fought in Vietnam. Nasatir engaged director John Irvin, a documentarian with experience in Vietnam, who noted, "All I can say is the film is a labor of love. It was made out of a great sense of compassion for the kids who fought there."[15] As Carabatsos noted when he was still trying to get *Hamburger Hill* made, "It's for the guys who were there, for their families. I'm hoping maybe some wife [of a veteran] will understand her husband a little better, or some kid will understand his father a little better."[16] *Three Kings's* (1999) writer and director, David O'Russell, was driven to make this Gulf War film by his sense for "veterans' mixed feelings about the end of the war."[17] Director Kathryn Bigelow and writer Mark Boal were motivated to make *The Hurt Locker* by a belief that the Iraq War had been underreported, and hoped to make what one journalist called "a character-based action movie [that] might give people of all political stripes a palpable understanding of life on the front lines."[18] When Bigelow won Best Director at the 2009 Academy Awards, she drew no attention to the moment as a historic first for a woman and instead dedicated the win to American soldiers, men and women, around the world, noting in closing, "May they come home safe." Her statement locates *The Hurt Locker* within historic discourses around the

PCF as a soldier-centered genre, although with the added dimension of a war film about a war still raging.

This rhetoric of moral urgency linked to action filmmaking dates to the earliest PCF to enter production: *Apocalypse Now*. (Finishing the film took so long that three other Vietnam films beat it to theaters.) Director Francis Ford Coppola pitched *Apocalypse Now* in this way to United Artists: "This is a high-quality action-adventure spectacle. . . . It's big and entertaining, mature and interesting."[19] In the press kit, Coppola articulated his goal "to put an audience through an experience—frightening but violent only in proportion with the idea being put across—that will hopefully change them in some small way."[20] And in his introduction for the printed program distributed at *Apocalypse Now*'s premiere showings in 70mm, Coppola stated, "It was my thought that if the American audience could look at the heart of what Vietnam was really like—what it looked like and felt like—then they would be only one small step away from putting it behind them."[21]

Coppola makes an astonishing claim for what a film can do in the public sphere: for him the experience of seeing *Apocalypse Now* could begin to heal the trauma of Vietnam. PCFs have mostly been exercises in catharsis and closure—an affective goal somewhat out of reach for twenty-first-century PCFs depicting ongoing wars in Iraq and Afghanistan. Steven Spielberg articulated a similar goal for *Saving Private Ryan* in a prerelease interview: "This isn't the kind of movie you see and then go to a bistro and break bread talking about it—you have to go home and deal with it privately. I think the audience leaves the theater with a little bit of what the veterans left that war with, just a fraction."[22] A published collection of online posts about the film on the still-new website America Online suggests that *Saving Private Ryan* worked in much the way Spielberg desired. Posts excerpted in the book *"Now You Know": Reactions after Seeing Saving Private Ryan* (1999) provide insight into the serious work PCFs can do for some viewers in the space of commercial entertainment.

- "[Spielberg] didn't use the tricks of the trade for cheap entertainment, but to help us transcend what we know of our lives."

- "I have never exited a movie theater in my 70 years of viewing movies where you walked in silence, holding back personal tears as you remember the past."

- "I am proud not only that I wept openly many times during the movie, but that my teenage son (a very tough acting kid) said, 'Anyone who doesn't cry at this movie isn't normal.'"

- "I hated war to begin with, but this movie made me have even more contempt for combat. I really believe if it were feasible, that if everyone on the face of the earth today could see this movie, there would be no more wars."

- "Do not dismiss this enlightenment as insubstantial because it's inspired by cinema. . . . This is what cinema is meant to do."[23]

As part of their discourse of catharsis and closure, many PCFs have sparked public conversations going beyond the entertainment press. As one cultural critic noted in 1979, "America is debating itself again on the Vietnam War. One movie has triggered this debate: *The Deer Hunter.*"[24] *American Sniper*, a film about the Iraq War made after the war had officially ended, followed much the same trajectory as *The Deer Hunter*, only in a new media environment. Endless discussions on cable news and the Internet turned *American Sniper* into an opportunity to re-prosecute the Iraq War—many analogous to discussions of the Vietnam War initiated by *The Deer Hunter*: both films present a white American warrior killing bloodthirsty foreign others. *Platoon* scored the covers of *Time* and *Newsweek* in articles about how the film presented Vietnam "as it really was." *Saving Private Ryan*, with dueling news magazine covers in the same week, elicited complementary media conversations: a pious discourse about the nation's debt to the so-called "Greatest Generation"; another about the effects of violence in film. It was generally seen as an appropriate use of graphic violence precisely because it served the purpose of educating viewers about the sacrifices of America's soldiers. Here, action-movie violence had a socially uplifting purpose. *Black Hawk Down*, opening in December 2001, emerged without intention as an interpretive football for the larger debate about how to proceed in the immediate post-9/11 era. Director Ridley Scott hoped *Black Hawk Down* would elicit a consideration of the dilemmas of intervention, a newly urgent topic.[25] Members of the group Act Now to Stop War and End Racism, profiled in the *Village Voice*, maintained that the film was "A conspiracy! A dangerous game of footsie between the Pentagon and Hollywood, created only to whet the country's appetite for more war." The article went on to note that this might have been the result of the protestors not having seen the film, which the paper read as "the ultimate FUBAR [World War II slang for "fucked up beyond all recognition," reintroduced to American audiences in *Saving Private Ryan*]. The viewer is more apt to leave the theater with a convincing impression that war is bad, war never works, and US troops should never be in Somalia again."[26] Conservative commentator Nicholas Kristof worried that *Black*

Hawk Down was "regrettably, a pretty good movie" that unfortunately taught "1. Nation-building is bloody, costly and futile," and "2. Casualties are completely unacceptable in American military operations."[27] Kristof feared that excessive caution in intervening abroad had created the conditions for 9/11. All these competing readings were served by the film's ambiguous approach to combat and its depiction of an American debacle.

Discourses of authenticity are central to the PCF. Most all these films purport to take the viewer onto the modern battlefield, and their means of doing so have created genre-specific practices, such as preproduction boot camps for actors, and special on-set creative figures, such as military advisors, whether independent or connected to the Pentagon. The Department of Defense has long participated in Hollywood's depiction of the military, and the PCF proves especially interesting in this regard. Unsurprisingly, many PCFs did not earn Pentagon approval—which can translate into access to military locations (bases), materiel (tanks, helicopters), and personnel (soldiers, pilots). But a good number of PCFs have won Pentagon support, suggesting—as Lawrence Suid has shown—that the US military understands that thematically complex combat films can serve their purposes as well as more one-dimensional movies—that, for example, *Black Hawk Down* or *Lone Survivor*, alike about failed missions, can be just as powerful cultural tools as a full-on fantasy like *Top Gun*.[28]

Another route toward prestige and authenticity involves foregrounding a film's access to real soldiers and veterans. Historian Stephen Ambrose's works of popular history, based on more than a thousand oral history interviews with veterans, fundamentally shaped Spielberg's combat output about Americans fighting in World War II, which encompasses directing (*Saving Private Ryan*), coproducing (*Flags of Our Fathers, Letters from Iwo Jima*), and executive producing for cable television (*Band of Brothers, The Pacific* [2010]). Ambrose's influence shows up quite literally in the scripts for *Saving Private Ryan*. Robert Rodat's original script was a rough-and-ready adventure flick with a happy ending. By the final shooting script, Spielberg had infused the project with Ambrose's view of the war, summed up in a prescriptive text laying out the intended larger interpretation of on-screen events: "*We are watching the true miracle of D-Day taking place*: when all the planning failed, when all of the calculations proved wrong, when the whole damn thing fell on its ass . . . *it was the common soldier who made it work anyway*. They seized the day in dribs and drabs, desperately improvising their way to victory in small rag-tag groups like this one."[29] Ambrose, the film's historical consultant, stated unequivocally in the press book that, while the narrative about saving Private James Ryan was fiction, "The film catches what happened exactly. It is, without question, the most accurate and realistic depiction of war on screen that I have ever seen, not only in terms of the action, but the actors look, act, talk, walk, bitch, argue and love one another exactly as the GIs did in 1944."[30]

Late in his 1997 book *Citizen Soldiers*, Ambrose briefly mentions a group of recent former soldiers who befriended him during his junior high years in Whitewater, Wisconsin. He recalls their scarred bodies seen during shirts-and-skins games of basketball and overnight hunting trips where they told him his first war stories. Writing some fifty years later, Ambrose confessed, "I've been listening ever since. I thought then that these guys were giants. I still do."[31] Ambrose's personal interaction with and sense of awe for the men who, in his view, won World War II was transferred to the actors during the making of *Band of Brothers*. Individual actors corresponded or spoke on the phone with the men they were playing. On one occasion, three surviving members of Easy Company visited the set. The *New York Times* described the scene, where "young actors were being called by the veterans' names," as "a surreal high school reunion without the name tags: older and younger selves meeting and exchanging suspicious but affectionate glances." In line with Ambrose's perspective, one actor characterized *Band of Brothers* as "about a type of man that's no longer created."[32] The theme of generational obligation resounded in the comments of actor Donnie Wahlberg: "I can safely say I speak for 98% of the guys in the show—this role was a two-year payback to the veterans of World War II."[33] Here, performance in a PCF is discursively framed as a faithful act of memorialization attended to by the actors themselves in a state of proper reverence. Media discourse around subsequent PCFs picked up this theme. Meet-the-veterans experiences circulated in the press around *Black Hawk Down*, reportedly shaping individual performances—Tom Sizemore said of his character Lieutenant Colonel Danny McKnight, "He didn't run or duck for cover [during the raid] because he didn't want to show the men under him that he was afraid. So my character doesn't, either"[34]—and creating a climate where "the actors were extremely aware of the need for them to be true to the men they were representing on the big screen."[35] *We Were Soldiers* director Randall Wallace solemnized that film's effort to "be true" by organizing "a service at the Fort Benning chapel for the survivors of the Ia Drang battle and for the cast and crew who would tell their story" the day before shooting commenced.[36]

The PCF strategy of connecting with real soldiers moved on-screen in the twenty-first century. *Black Hawk Down* enjoyed extraordinary cooperation from the military—an investment in men and materiel unlikely to have been available had the film gone into production just after instead of just before 9/11. Mark Bowden, author of the book *Black Hawk Down*, reported, "Most of the military stunts performed in the film, from flying choppers to roping Rangers, were performed by actual members of those army units—in some instances, soldiers who had fought in the battle [of Mogadishu] themselves."[37] The practice of using active-duty soldiers or recent veterans who were actually in a depicted conflict intensified in subsequent PCFs about the Iraq War. Recon Marine Rudy Reyes plays himself in *Generation Kill* (2008). Iraq War veterans in generally nonspeaking roles surround star Matt Damon in *Green Zone*. Damon noted, "The whole point of these guys being here is that they show up and are who they really are. That's not something that a group of actors, even with a long time to work, could pull off as well as a group of veterans."[38] The four Iraq veterans cast in *Green Zone* "came aboard as actors—mostly background performers— but also served as unofficial military consultants."[39] The quoted article framed the veterans' opportunity to replay their military selves in a film as therapeutic. As noted, *Act of Valor* cast actual Navy SEALs in leading and supporting roles. The SEALs, credited by first names only, were effectively ordered to perform. Their rather limited dramatic range constantly reminds the viewer that these are not actors. But while they might lack skill as actors, actual military men do not bring established personas to their parts. This difference comes into focus late in *American Sniper*, when SEAL sniper Chris Kyle—played by Bradley Cooper—listens to a group of wounded veterans discuss their combat experiences. It's hard to register Cooper as Kyle—and not a handsome movie star—in such a context.[40] *Lone Survivor* offers a strange case where a narrative of memorialization incorporates the cinematic sacrifice of the survivor who lived to tell the tale. Marcus Luttrell, the SEAL who wrote the book, appears several times in the film, moving through the story like a ghost, his presence legible for those who know Luttrell's face from the photos in his book or from the media. He eats breakfast with the main characters—including actor Mark Wahlberg, who is playing Luttrell—early in the film and delivers his one line. He's also present at the new-guy ceremony, a key formal transition in the narrative. And Luttrell is on the Chinook helicopter that is shot down with a single RPG (rocket-propelled grenade), an abrupt and emotional moment of loss in the story. The real Luttrell dies symbolically on film with the actors playing the men who died trying to save him in Afghanistan—men whose real faces are seen during the memorializing musical sequence at the film's close. In the above cases, PCFs offer the viewer the opportunity to behold the actual bodies of military men in action or in symbolic performance.

The goal of authenticity also generally unifies PCFs' visual style. Unlike the enhanced modes of action-adventure cinema discussed by Lisa Purse in *Contemporary Action Cinema* (2011)—where spectators regularly accept and cheer acts of physical prowess only possible in the movies—the rules of the physical world remain in operation over the PCF, which has frequently been marketed under the fraught term *realistic*.[41] Several directors have self-consciously limited their visual vocabularies when making PCFs. Spielberg (as executive producer of *Band of Brothers*) and Sam Mendes (on *Jarhead*) both set aside the use of crane shots, keeping the camera at eye level, constraining the storytelling to a human point of view.[42] Low-budget PCFs of necessity stick to the ground. Action directors with strong visual styles have had difficulty meeting the authenticity demands of the PCF. Woo's *Windtalkers* was criticized for this. Said one critic, "His multicamera slo-mo balletics don't really conjure up the heat of battle; they conjure up other John Woo movies."[43]

The PCF trades on a supposedly invisible, "realistic" visual style—itself a historical construct that changes across the four-decade history of the subgenre. For example, twenty-first-century PCFs draw on multi-camera, handheld documentary film to make this claim. In *Platoon* it was enough to have mud splash on the lens during the opening credits, suggesting a visceral presence for the camera. The special effects arsenal that has increasingly defined Hollywood film over the last four decades is turned to specific ends in the PCF. Purse's discussion of *United 93* applies broadly to the subgenre: "The digital visual effects function as a solution to a number of practical challenges in order to help maintain the sense of cameras capturing events 'as they happen' in a naturalistic, realistic-looking environment. That is, digital imaging interventions allow the [film] to produce the illusion of photographic indexicality."[44] This use of digital effects works at various levels of scale, from simulations of the massed machines of war—breathtaking images of Allied armadas in the water off Iwo Jima in *Flags of Our Fathers* or in the air above Germany in *Red Tails* (2012)—to startlingly intimate moments where characters bleed to death before the viewer's eyes, depicting the effect of bullets on the human body at a graphic level that often leads to an R rather than PG-13 rating. (The sucking chest wound runs across the subgenre as a recurring, always increasingly graphic, bodily trope.) PCFs are, almost by definition, not intended for the twelve-to-seventeen-year-old action cinema demographic. Made for (hopefully) more reflective adults, in these films gravity works, flesh fails, and death matters.

Philip Drake has described the "peculiarities of Hollywood films as cultural goods," noting that the ticket price is the same for all films, and that product differentiation works differently than with other goods. In the case of films during their theatrical release periods, the price of "admission might buy the social experience of cinemagoing rather than to see a particular film."[45] PCFs offer an experience in the cinema that activates moviegoers' sense of themselves as citizens—and, for some, as soldiers (real or prospective) and veterans. The social experience of seeing a PCF has often been profound and, again unusual for action films, has at times cut across the four "quads" that define Hollywood marketing: male/female, under twenty-five/over twenty-five. The multigenerational family pictured in the opening and closing of *Saving Private Ryan* directly mirrors the desired audience for the film itself. (Chapter 3 of this book considers how this priority has shaped PCF dialogue, often moderating crude language presumably so as not to alienate women or older audiences.) Still, PCFs reach out to young men in particular.

Charles Acland writes, "Motion pictures have a life-cycle through which their cultural and economic impact rises and falls" made up of the "range of media forms through which a cultural text travels."[46] Many PCFs have had long lives, shaping how generations of young men think about war and soldiering. The subgenre's impact was amplified by the arrival of the VCR in the 1980s. Iraq War veteran Colby Buzzell notes in his 2005 memoir *My War*, "[My generation] grew up watching [movies like *Apocalypse Now*, *Full Metal Jacket*, *Platoon*, *Hamburger Hill*, and *Black Hawk Down*] over and over again and can recite word for word countless lines from each, and most of us were probably here in the Army because we watched these movies one too many times."[47] The PCF is a part of popular military culture, even as the subgenre's makers continually adjust its representation of the military. The Recon Marines profiled in the 2004 book *Generation Kill* screened *Black Hawk Down* together before deployment and quote to each other from *Platoon*.[48] In his 2003 memoir *Jarhead*, Gulf War veteran Anthony Swofford recalls the pre-deployment "Vietnam War Film Fest" he shared with his Marine buddies and offers a warning of sorts about the various ways these films might be interpreted:

> There is talk that many Vietnam films are antiwar, that the message is war is inhumane and look what happens when you train young American men to fight and kill, they turn their fighting and killing everywhere, they ignore their targets and desecrate the entire country, shooting fully automatic, forgetting they were trained to aim. But actually, Vietnam war films are all pro-war, no matter what the supposed message, what Kubrick or Coppola or Stone intended. Mr. and Mrs. Johnson in Omaha or San Francisco or Manhattan will watch the

films and weep and decide once and for all that war is inhumane and terrible, and they will tell their friends at church and their family this, but Corporal Johnson at Camp Pendleton and Sergeant Johnson at Travis Air Force Base and Seaman Johnson at Coronado Naval Stations and Spec 4 Johnson at Fort Bragg and Lance Corporal Swofford at Twentynine Palms Marine Corps Base watch the same films and are excited by them, because the magic brutality of the films celebrates the terrible and despicable beauty of their fighting skills. Fight, rape, war, pillage, burn. Filmic images of death and carnage are pornography for the military man; with film you are stroking his cock, tickling his balls with the pink feather of history, getting him ready for his real First Fuck. It doesn't matter how many Mr. and Mrs. Johnsons are antiwar— the actual killers who know how to use the weapons are not.[49]

Swofford reminds us of the diversity of ways a single film might be seen, and the inherent ambiguity in anything we might call a "serious" war film.

In his 2008 book *War and Film*, James Chapman usefully distinguishes didactic war films, which leave no room for contested readings, from others that are more open to varied readings and responses.[50] PCFs are of the latter type, and music, as will be shown, plays a large role in securing such openness. Still, Chapman warns that either type of war film can be seen as part of what Graham Dawson calls the "pleasure culture of war."[51] Make no mistake, the PCF—a genre oriented in almost every individual case toward a broad commercial market—sits squarely within the pleasure culture of war. Again, music—even if played in response to tragic, meaningless battlefield losses—is part of the pleasure of that culture.

THE FOUR CYCLES OF THE PRESTIGE COMBAT FILM

The PCF subgenre unfolded in four topical cycles of varying intensity and duration, some overlapping chronologically. Three of these cycles—depicting World War II, Vietnam, and US wars of the 1990s—were retrospective in nature, looking back at conflicts that were over, typifying what the historical film scholar Jonathan Stubbs has called "small-scale, historically specific film cycles which emerge from particular commercial contexts and are shaped by larger cultural forces."[52] The fourth PCF cycle, depicting US wars in the Middle East in the twenty-first century while those wars were ongoing, functions somewhat differently. This especially fraught cycle required filmmakers to respond to a deeply divided national mood and to represent profound changes in the nature and scope of soldiering. How each cycle relates to the figure of the soldier and the veteran is sketched in brief below. Understanding these four cycles proves important, as expressive and structural sonic strategies discussed in later chapters will often be situated as either typical of a given cycle or as carrying across cycles.

The Vietnam cycle is foundational to all PCFs that follow. Initiated on the production end by *Apocalypse Now*, these ten films collectively defined the PCF as both an exciting action-driven genre with combat scenes and a serious meditation on soldiers and veterans. With its primary focus on the depiction of American soldiers in the thick of combat, the Vietnam PCF embraced both epic, expensive, elite moviemaking and modestly budgeted but thematically earnest films. The subgenre was effectively open to all levels of budget from the start, and most films reached out quite directly to the veteran audience, often in dedicatory titles. While two signature Vietnam PCFs were made by top directors able to set their own course (Coppola's *Apocalypse Now*, Stanley Kubrick's *Full Metal Jacket*), most were either low-budget independent productions (*Go Tell the Spartans* [1978], *84 Charlie MoPic* [1989], *Hamburger Hill*) or foreign-financed films the Hollywood studios initially would not touch (*The Boys in Company C* [1978], *The Deer Hunter*, *Platoon*).[53] Only after the success of *Platoon* did one major studio—Columbia—back a final pair of Vietnam PCFs: Stone's *Born on the Fourth of July* (1989, made on a grand scale as a follow-up to the decidedly smaller *Platoon*) and Brian De Palma's *Casualties of War* (1989, facilitated by the director's recent hit *The Untouchables* [1987] and the willingness of new Columbia studio president Dawn Steel to fund, in her words, "some movies we have a passion for, respect for").[54] Most all the makers of these films were either directly connected to the Vietnam War—more than a few served in or reported on the war—or lived through the Vietnam era as adults. The 1980s cycle almost completely exhausted Hollywood's interest in Vietnam. Only *We Were Soldiers* has been made since 1989.

The early 1990s saw a lull in PCF production even as US military engagements abroad created new potential subjects for combat films. Hollywood eventually produced three films (*Courage Under Fire, Three Kings, Jarhead*) about the Gulf War—the 1990–91 mission, known as Operation Desert Storm, to remove the Iraqi army under Saddam Hussein from Kuwait—and one film (*Black Hawk Down*) about the 1993 mission to Somalia, which came to grief on the streets of Mogadishu. The Gulf War films form an odd, chronologically scattered trio that do not cohere beyond their similar historical setting. These films do not reference each other, nor do they explore the high-tech, mediatized nature of the Gulf War. Instead, each takes up questions closer to the ground and more germane to the issues of soldiers and veterans at the heart of the PCF. *Courage Under*

Fire—written by Vietnam veteran Patrick Sheane Duncan, writer-director of *84 Charlie MoPic*—is not in the slightest about the Gulf War. Instead, the film, which one critic described as "essentially Vietnam movie maneuvers in sand-colored fatigues," offers familiar Vietnam tropes: downed choppers, an unseen enemy advancing on an isolated position in the night, the fragging of officers, incidents of friendly fire, and cover ups by the military.[55] The Pentagon refused to cooperate with the production on the grounds that the mutiny at the center of the plot would be "astonishing behavior for the all-volunteer, post-Vietnam Army," further evidence that the film is, in fact, about the Vietnam-era draftee army.[56] The original *Three Kings* screenplay, titled *Spoils of War*, was a straight-ahead action flick, a revenge plot modeled on *The Treasure of the Sierra Madre* (1948).[57] As the *New York Times* reported, in his revisions "the more politically motivated [writer and director David O. Russell] took aim at the cultural stew that spilled American consumer goods all over the desert while everybody tried to determine the point of the war. To him the point of the movie was a failure by the United States to support large segments of the Iraqi population."[58] Using the PCF to critique specific military and foreign policies has been a very rare approach: the only later film to do so pointedly is director Paul Greengrass's *Green Zone*. The third Gulf War PCF was made during the Iraq War. Director Sam Mendes's 2005 film *Jarhead* is based on veteran Anthony Swofford's best-selling memoir from 2003: both were out of phase with unfolding history. By the time Swofford's tale was told in print and on film, it was a period piece. The film's characters show intense frustration that they never get to experience real combat in the hundred-hour Gulf War, and audiences were expected to process this dilemma while (perhaps) following intense and bloody battles in the Iraq War, at its height when *Jarhead* was released to middling business. And as time passes, *Jarhead* feels more and more out of sync. Its plot turns on a pair of scout snipers who never take a shot. *American Sniper* tells the story of a Navy SEAL sniper with the most confirmed kills in American military history— a distinction facilitated by ten years of sustained combat in Iraq. Still, *Jarhead* offers a useful contrast to other post-9/11 depictions of the all-volunteer military. The soldiers in *Jarhead* are misfits, described by an officer as "some weird motherfuckers." This jars strongly against twenty-first-century trends to present contemporary soldiers in the all-volunteer force as calm, competent professionals.

Steven Spielberg's *Saving Private Ryan* and Terrence Malick's *The Thin Red Line* were released within months of each other in 1998 and together launched a long-running cycle of World War II films. (Curiously, the

American soldier's experience in World War I has gone entirely unrepresented in post-Vietnam cinema.) As Michael Hammond has pointed out, both are "special cases. . . . prestige projects the green light for which was made possible only through the reputation of the film-makers themselves."[59] Both continue, on much less controversial ground, the exceptional practice of *Apocalypse Now*, described by Walter Murch as "a personal film, despite the large budget and the vast canvas of the subject."[60] In the wake of these 1998 entries, the topic of World War II attracted directors Clint Eastwood (*Flags of Our Fathers*, *Letters from Iwo Jima*), John Woo (*Windtalkers*), and Spike Lee (*Miracle at St. Anna* [2008]), as well as producer George Lucas (*Red Tails*) to the PCF. The final three titles, in different ways and to shared failure at the box office, endeavored to insert Native American and African American veterans' stories into Hollywood narratives of World War II, suggesting how PCFs might address the unspoken whiteness informing most films in this cycle. (The interracial dynamics of the Vietnam and Gulf War cycles prove an important aspect of chapter 4's discussion of popular music in the PCF.) The decidedly noncommercial *Letters from Iwo Jima*, perhaps the only entirely foreign-language film ever made by a major Hollywood director and studio, is among the most remarkable members of the subgenre—a testament to the ability of major auteurs to create unusually challenging work within the studio system when the topic is American soldiers and veterans. *Letters from Iwo Jima* complicates the memory of World War II for American audiences by drawing the viewer into the varied experiences of Japanese soldiers.

Saving Private Ryan spawned two limited series on HBO—*Band of Brothers* and *The Pacific*—both executive produced on a cinematic scale by Spielberg and actor Tom Hanks, their shared star personas shaped fundamentally by *Saving Private Ryan*'s stature as a cultural event. These limited series—together with the 2008 HBO series *Generation Kill*, about the 2003 invasion of Iraq—share much in the way of narrative, formal, and musical tropes with the feature films at the core of this study. Given the way most Americans watch new (and old) movies in the twenty-first century—at home on a television or on some sort of computer screen—inclusion of these prestige combat cable series in this study makes sense.

As noted above, *Saving Private Ryan*, like *Three Kings*, began as a standard action script that was substantially rewritten by the director into a more ambiguous film. The addition of the old Ryan in the framing scenes set in the present brought this World War II narrative into line with the veteran-centered Vietnam cycle, with the difference that the burden is generational. Indeed, almost the entire World War II cycle is driven by an intergenerational dynamic, with mostly baby-boomer filmmakers saluting their fathers. All the World War II PCFs—with the notable exceptions of *The Thin Red Line* and *Letters from Iwo Jima*—more or less answer the perennial question "Grandpa, what did you do in the war?" in a complex but ultimately affirmative manner. The American men who fought are revealed as gentle souls, intent on staying alive and supporting their buddies in the hope of making it back home. These are Ambrose's citizen soldiers who long for peace—not the conflicted draftees of Vietnam, nor the skilled professionals of the Gulf and Iraq War films.[61]

Films made before but released after the terrorist attacks of September 11, 2001, fall into an uneasy reception context. Indeed, the Hollywood studios showed some trepidation about war films in the immediate aftermath of 9/11. For example HBO abruptly discontinued its huge promotional campaign for *Band of Brothers*, the premiere episode of which had aired on September 9. But a noticeable spike in patriotism encouraged accelerated release of *Black Hawk Down* (from March 2002 to December 2001) and *We Were Soldiers* (from summer to March 2002). Of the latter, Paramount chairwoman Sherry Lansing said, "It's about the sacrifices that solders make so the rest of us can be safe. I think we're ready for that at any time, but now it's particularly relevant."[62] *Black Hawk Down*, presenting urban combat in the Middle East, would prove an important model for the fourth PCF cycle's depiction of twenty-first-century American soldiers in Iraq, Afghanistan, and elsewhere across the globe.

The so-called Global War on Terror (or GWOT), initiated in no uncertain terms by President George W. Bush shortly after 9/11, created new challenges for filmmakers. Unlike the cinematic reaction to Vietnam, which lagged behind the prosecution of the war at a safe distance, Hollywood filmmakers responded to the GWOT while the conflict was going on, in spite of the fact that American audiences showed little to no interest in films about the war—or, by some accounts, in the ongoing war itself. Still, the GWOT PCF beckoned, creating the possibility of making what director Kathryn Bigelow called the "Holy Grail of filmmaking," defined by the *Los Angeles Times* as "an entertaining genre movie that opens a window into a current event."[63]

Two successive subcycles of GWOT PCFs dealt differently with the war, with contrasting results at the box office and with the critics. The first subcycle, from 2006 to about 2010, told tales of American military frustration and was marked by tremendous generic innovation and pervasive commercial failure. In a passing 2008 reference, *Variety* christened these films a "toxic genre."[64] The second subcycle, turning toward the elite Navy SEALs

in the early 2010s, proved more successful commercially while also generating controversy and conversation in the culture at large.

With the exception of the special case of *United 93,* most combat filmmakers in the first GWOT cycle gravitated to the conventional ground war
in Iraq, which proved most amenable to existing conventions of the combat
film. *Black Hawk Down* provided a model to follow or work against for
PCFs set in the urban Middle East such as *Green Zone, The Kingdom*
(2007), *Redacted* (2007), *Generation Kill,* and *The Hurt Locker.* (The earliest Iraq War combat film, director Sidney J. Furie's independently produced,
Canadian-financed *American Soldiers* [2005], was never picked up for theatrical release in the United States but is available on DVD. Furie directed
The Boys in Company C, the first of the Vietnam PCFs to be released.)

Green Zone, The Kingdom, and *Redacted* typify the twenty-first-century PCF as a generic hybrid. Paul Greengrass's *Green Zone* explores the
search for weapons of mass destruction (WMDs) in post-invasion Iraq by
way of a Jason Bourne–style "propulsive, paranoid conspiracy thriller" (as
characterized by Universal cochair Donna Langley).[65] Peter Berg's *The
Kingdom* crosses the combat genre with the procedural, an ascendant genre
in American television at the time, to tell a fictional story of FBI investigators who travel to Saudi Arabia to investigate a massive terrorist bombing
of a compound housing American civilian oil company workers. *Green
Zone* and *The Kingdom* alike followed *Black Hawk Down*'s innovative continuous-action style that would come to mark much of action cinema in the
twenty-first century. The highly controversial *Redacted* by writer-director
Brian De Palma—shot in Jordan in two and a half weeks on high-definition
video for $5 million and released on just fifteen screens in the United States
despite De Palma winning the Silver Lion for best director at the Venice
Film Festival—uses a collage of documentary film styles to tell a tale of
American soldiers raping and murdering Iraqi civilians. Both De Palma's
PCFs—*Redacted* and *Casualties of War*—reveal a director using the feature film to highlight the potential of American fighting men to perpetrate
atrocities, specifically the rape and murder of young women. The very different financing for these two thematically similar films sheds light on the
director-driven nature of the subgenre and the varying industrial potential
for the PCF to cast "structured doubt on the innocence of American soldiers," a potentially disturbing and unusual extension of the subgenre's
expressive remit.[66]

The GWOT dramatically enlarged the potential PCF cast of characters,
as intelligence services and domestic agencies, even perhaps civilians, were
all involved in the fight. As noted, *The Kingdom* centers on FBI investiga-

tors. *United 93* dramatizes the difficult communications situation between civilian air traffic control and military command on September 11. Indeed, *United 93* sits on the edge of the subgenre, just as the actions of the crew and passengers on United flight 93 have been absorbed into narratives of the GWOT. (Having learned of successful attacks on the World Trade Center and the Pentagon, they resisted their hijackers and crashed the plane outside Shanksville, Pennsylvania.) On the first anniversary of 9/11, General Tommy Franks called the events on that plane "the first battle on terrorism."[67] As such, *United 93* shares with other PCFs the urge to use commercial cinema to memorialize those who lost their lives in service to the nation by re-creating scenes of combat, offering moviegoers the chance to leverage a movie ticket as an act of patriotic remembrance and sympathetic understanding extended toward America's soldiers, in this unique case civilians anointed as such after the fact.

Among PCFs of any cycle, *In the Valley of Elah* (2007) includes the least amount of combat action. Written and directed by Paul Haggis, it is based on the true story of an Army unit deployed to Iraq who, on their return to the United States, murdered a soldier among their number, then attempted to conceal the crime by chopping his body into pieces and burning the remains.[68] Told through the murdered soldier's veteran father's search for the truth, the procedural-like narrative contains a second, combat-centered line. The father hires a computer tech to recover the video and photo files on his son's damaged cell phone. The restored files give the father (and the viewer) glimpses of combat in Iraq, including the abuse of prisoners, and children killed by American soldiers driving without stopping so as to avoid roadside bombs and ambushes. The conceit of authenticity here—as in *84 Charlie MoPic* and *Redacted*—rests on the notion of found footage. The cultural work of the PCF centers on bringing home the experience of combat in a given war. *In the Valley of Elah*, without a doubt, does this—to troubling ends. The film closes with the image of an American flag flown upside down, a signal of existential distress.

The standout film in the first GWOT cycle, *The Hurt Locker*, centers on explosive ordnance demolition (EOD) teams tasked with defusing the signature weapon of the Iraq War: improvised explosive devices (IEDs). It was highly praised by critics as an innovative action film cutting to the heart of the unique nature of the Iraq War, its central character capturing the first GWOT subcycle's ambivalence toward the contemporary soldier. Staff Sergeant Will James, played by Jeremy Renner, is a skilled EOD tech tasked with defusing one IED after another, with no end in sight. His emotional opacity and evident satisfaction with this life—James confesses it has become the only thing he loves—indirectly raises uncomfortable questions about the US military and militarism. The film's epigraph—"The rush of battle is often a potent and lethal addiction, for war is a drug"—comes close to articulating this theme in no uncertain terms. This combat mise-en-scène found few willing moviegoers. For all its critical acclaim, *The Hurt Locker* remains the lowest-grossing Best Picture winner in history. Renewal of the PCF in the era of the GWOT required drafting a different professional soldier to embody the ongoing battle in more conventionally heroic and cinematic ways.

The second GWOT cycle kicked off in 2012 with *Act of Valor*, an innovative PCF starring actual Navy SEALs. In his foreword to the novelization of *Act of Valor*, the best-selling military thriller author Tom Clancy states, "Navy SEALs are Olympic athletes that kill people for a living."[69] In other words, the SEALs are natural and authentic action movie stars. Their ascendance in Hollywood combat films reflects a recalibration of PCF narratives to match the nature of the GWOT as prosecuted by the Joint Special Operations Command (JSOC), which unites Special Forces from all branches of the military and conducts operations across the globe via a mix of high-tech drone warfare and smash-and-grab raids.[70] The journalist Peter Bergen notes, "In the decade after 9/11, JSOC mushroomed from a force of eighteen hundred to four thousand, becoming a small army within the military."[71] With the end of conventional force commitments in Iraq and Afghanistan, JSOC missions continue without abatement. The second GWOT subcycle has drafted the SEALs as Hollywood stand-ins for the entire US military. And while drawing directly on earlier, more ambiguous PCFs about American soldiers overseas, the SEAL Team PCFs have surely contributed to what one critic of the military calls "the semimythic Special Operations Command," which promotes the efficacy of "heavily publicized 'secret' warriors" pursuing high-value targets.[72]

Black Hawk Down, centering on a combined mission involving two elite units—Army Rangers and Delta Force—serves as a clear precursor to the SEAL Team cycle. One youthful Ranger even counsels his men, "We're elite—let's act like it, " a line unlikely to be heard spoken by one of the seasoned, laconic warriors in the SEAL GWOT cycle. One critic described this sort of soldier as "a world-class expert—superbly trained, heedlessly brave, a figure set very much apart from the rest of us," and by extension fundamentally different from draftees of Vietnam and the citizen soldiers of World War II.[73] Here, Hollywood found heroic, real-life figures whose "sheer professionalism" could obviate any narrative of military failure.[74]

The soldier as professional—cool, controlled, competent—dominates the depiction of SEALs. *Act of Valor* initiates this characterization perhaps of necessity: using real SEALs as leading men limited the emotional range of the characters; music steps in and does heavy work, as shown here in subsequent chapters. The SEALs enter *Zero Dark Thirty* late in the film, after CIA analysts have found Osama bin Laden's compound in Abbottabad, Pakistan. Passion and sacrifice are on the side of the analysts, whom Bigelow described as "soldiers of a different type, right? And they're not in uniform, and they're not on the front lines, but they're warriors, I guess, is the word I'd use."[75] The expanded cast of *Zero Dark Thirty* expresses what William M. Arkin has called "our wholly transformed hybrid of a military."[76] The SEALs who execute the night mission to kill bin Laden in the film's final reels show minimal emotion.[77]

Lone Survivor and *American Sniper*, alike based on memoirs by Navy SEALs, of necessity invest in the emotional lives of the professional warriors at their center. Both films do this while simultaneously departing significantly from the thematic thrust of their respective sources. Neither film adopts the tone of the books, which are, in large part, conservative diatribes against vaguely defined liberals in government and the media. While clearly calibrated to not alienate supporters of the GWOT—and thereby more conservative in their politics than most PCFs—the SEAL Team cycle is sufficiently nuanced to allow resistant readings while advancing the fundamental underlying theme of almost all PCFs: the sacrifice of soldiers for the nation memorialized in a narrative feature film.[78] As Clancy's foreword to *Act of Valor* the novel plainly states, "We have an obligation to honor the SEALs and their families."[79] Sitting through any of these PCFs—except perhaps *Zero Dark Thirty*, which casts the SEALs in walk-on parts—qualifies as meeting such an obligation, a rather low standard but of a piece with the distant relationship between the military and most American citizens in the era of the all-volunteer force.[80] The SEAL Team subcycle, on the whole very successful at the box office, effectively transferred the memorializing functions of the Vietnam and World War II PCFs to the very different context of the GWOT.

While action films, Hollywood's bread and butter since the late 1970s, are typically not morally complex, prestige combat films are, or aim to be. The entire subgenre can be grouped under a Hollywood oxymoron: ambiguous action movies. This flows in part from their story context and content. As the critic Andrew Sarris noted, "The war film is the one cinematic genre that can exploit massively homicidal violence while professing to make a moral statement about it."[81] But forged as the subgenre was in a delayed cinematic reaction to Vietnam, the PCF can push such "moral statements" into new territory, where the fact of American defeat must be accounted for and where the actions of the US government and military can be opened for debate. However, as the above survey shows, most PCFs dwell on the soldier and the veteran, filtering any larger questions through the experiences of individual characters with whom the movie audience can identify, and allowing viewers to selectively read these films. Young men can simply read them as pro-war.

The PCF demonstrates how, in the combat film scholar Jeanine Basinger's words, "different wars inspire different genres." Her terse formula "genre is alive" finds support again and again in the history of the PCF.[82] The remainder of this book shows how the PCF has been especially "alive" in the domains of music and sound, where innovative expressive tropes from the 1980s Vietnam cycle find different meanings in the World War II, Gulf War, and GWOT cycles. Along the way, questions of patriotism (love of country) and humanism (appreciation for the value of all human life, including the enemy) are constantly under negotiation. The ambiguous nature of the PCF opens a perhaps unlikely commercial space for these concepts to be explored. Mikkel Bruun Zangenberg, discussing Eastwood's Iwo Jima films, notes how "the Western umbrella terms 'humanism' and 'patriotism' . . . do not denote sharply delimited fields of meaning" but are instead "derived from a huge, if fuzzy, culturally saturated terrain of ideas, notions, beliefs, norms, and values." Zangenberg expresses concern that in Eastwood's two films these terms are only resolved in ambiguity. In fact, the capacity for ambiguity constitutes the aesthetic advantage of the PCF. As will be shown, music—the least prescriptive of the arts in its meanings—often powerfully serves such ambiguous ends. Still, as Zangenberg notes, there is for war films a *"strong, hermeneutical desire*, the desire that warfare not be meaningless, absurd, and futile."[83] Meeting this desire with *some* positive response—however equivocal—underpins the PCF in every instance (except, perhaps, for Kubrick's *Full Metal Jacket*).

Most of the time, of course, the response sought by PCFs is one of thankfulness for the sacrifice of the fallen, the posture proper to ritual acts of memorialization. To be surprised at this is naive, but it is hoped that the analysis of PCFs to follow shows how equivocal and unsure these films have been, at times, about such conclusions. I take issue with Vibeke Schou Tjalve's assessment that "though war politics after Vietnam has done a better job at welcoming veterans, its ability to come to terms with war—to mourn, reflect, and regret it—has not improved. A genuine public language of the tragic nature of all warfare remains absent."[84] In the realm of the PCF, and especially on the soundtrack, a new, modern language of the tragic nature of war can be heard. But, to reiterate, that language is relentlessly personal—it sounds at the level of the individual. And here, a change in military policy toward individual soldiers helps bring the nature of the PCF as a therapeutic subgenre into focus.

The poster tagline for *Black Hawk Down* reads, "Leave No Man Behind." A track on composer Hans Zimmer's score album for the film identifies a musical theme with the phrase. The military imperative to "leave no one behind" on the battlefield is of relatively recent vintage. Beth Bailey's *America's Army* (2009), a history of the all-volunteer force, locates formal articulation of "leave no man behind" in the so-called "warrior ethos," a rebranding of the post-9/11 Army that put the phrase "I will never leave a fallen comrade" into the Army Soldier's Creed as spoken ritualistically and as emblematized on special dog tags.[85] Leonard Wong traces the policy to Vietnam, when "search and rescue began to replace all missions as the most critical mission, . . . the tactical expression of the US strategy to bring home its troops." Wong quotes a rescued pilot who understood that if a man was lost, the official attitude was, "Okay, we're going to stop the war and get this guy back, and then we'll resume."[86] Wong also details specific post-Vietnam missions—such as Mogadishu in 1993—where engagements with the enemy turned into rescue operations to recover wounded or dead comrades in which further American lives were lost. Writing in a military policy journal, Wong questions the "rational sense" of a military ethic that demands soldiers' bodies always be recovered, noting that such a policy ends up becoming the mission of the military. In the context of commercial narrative cinema, this relatively recent combat prioritization opens the door to a certain kind of plot where—to quote the tagline for *Saving Private Ryan*—"This time the mission is a man."[87] Recovery of the wounded and the dead takes on a sacred quality, the only thing of importance in a generic discursive context where political discussions—Why are we there at all?—are tacitly ruled inadmissible. All that matters is recovery of the body, dead or alive—but hopefully the latter. The tools of cinema are great at telling this story.

War film scholarship at times seems to ignore this function of the combat genre, instead faulting these films for supporting the very idea of nation-states making war and using young men to do so. To quote a blog post by Slavoj Žižek: "However, we should bear in mind that the terse-realistic presentation of the absurdities of war in *The Hurt Locker* obfuscates and thus makes acceptable the fact that its heroes are doing exactly the same job as the heroes of *The Green Berets*. In its very invisibility, ideology is here, more than ever: we are there, with our boys, identifying with their fear and anguish instead of questioning what they are doing there."[88] Criticizing war films for not asking the "why" question of a given war makes them an easy target; indeed, not a target at all but instead a straw man. More interesting questions are, What are the moviegoers doing there? Why do audiences go to these films, which are more complex than the average action-adventure film? Why are the filmmakers there? What function does making the film serve them? And what is the substance of the "there" of these films, which are poised so delicately between genre conventions, conceits of authenticity, and the need to memorialize American men fighting and dying on foreign battlefields, whether for glory or for a lie, depending on the war? *Hymns for the Fallen* takes up these film-centered questions by listening to the PCF subgenre with an ear for how film form and musical form work together, and for how the sonic space of the soundtrack is mixed for narrative and ideological ends, producing over decades of creative ferment a group of cinematic war memorials that represent the figures of the soldier and the veteran in the post-Vietnam era.

2. Soundtracks and Scores

This book attends to the sonic content of prestige combat films (PCFs) by parsing their soundtracks along three expressive domains—dialogue, sound effects, and music—which in turn reflect discrete industrial and creative contexts. "Dialogue" refers to words meant to be understood and is primarily recorded on the set, then re-recorded (or looped) as needed in postproduction. Sound effects are central to combat film storytelling. The sounds of weapons and machines drive battle plots, organize the space of the battlefield, and bring particular war materiel to life—the rocket-propelled grenade (RPG) and helicopter have both been exploited for their cinematic possibilities as sound and image and as sound only. Sound effects professionals produce this element, frequently thinking of their work in musical terms. Music is essential to the subgenre—to a greater degree than in pre-Vietnam war films—and composers and music supervisors play important creative roles. Many PCFs are, by their very nature, highly musical films, where music plays a central role not only as expression but also as a constituent formal element, with film form often following musical form. Musical form and content prove essential to the meaning of almost all PCFs. In this sense, PCFs are quite literally hymns for the fallen.

Of the three elements of film sound, music (or the musical score) enjoys a certain autonomy. Scores are frequently experienced on their own, chiefly by way of the so-called original soundtrack album (which misapplies the term *soundtrack* to the musical score alone). For decades, albums containing just the musical score for a film—which I call score albums—have given audiences access to film music and film composers' versions of a given score in an adjacent, wholly independent format. Sometimes score albums prove revealingly different from the score as heard in the film, a situation where the album's tracks and the film's cues don't match. For example, composer

James Horner's score album for *Courage Under Fire* (1996) includes musically rounded, concert hall–ready renditions of the score's main themes in a manner unlike the way these themes are heard in the film. In addition, a long continuous track on the album is heard only partially in the film. In the mix, a good sixty seconds of music in the middle of this long cue are faded out completely to make way for important dialogue. The missing music can be heard on the album, which emerges on close listening as a kind of archival trace of the film that didn't get made.

Original scores usually suggest a creatively autonomous figure: the composer. It is possible to speak meaningfully of both Steven Spielberg's and John Williams's *Saving Private Ryan* (1998). But assessing film score authorship can be a tricky proposition. For example, *Apocalypse Now's* (1979) score was the work of many hands. Carmine Coppola, credited alongside his son director Francis Ford Coppola as composer, wrote several melodic themes and played flute. Four synthesizer players were brought in to add unique sounds drawn from their own custom-built synthesizers. Improvisers from the world of rock and roll—among them drummer Mickey Hart of the Grateful Dead—played along with the edited film, producing a parallel album of music purportedly (but not all that easily) heard in the film.[1] Various remixers and pianists worked on a practical level to combine all these sources, and Walter Murch, stepping from the editing to the sound department, supervised the mix. Music producer David Rubinson even remixed all three elements of the soundtrack into a double album—a radio drama of sorts, substantially different in its content from the film, a soundtrack album but not an original soundtrack album. All of these creative individuals worked to fulfill Coppola's "musical vision, which was inextricably tied to his overall view of the film."[2] Arguably, in this instance, the director is metaphorically—if not literally—the orchestrator of the score.

The soundtrack for *The Hurt Locker* (2008) was assembled in analogous fashion, with rich cross-fertilization between the sound and music departments and a goal of continuity across domains, although with less directorial input. Paul Ottosson, credited on the film as "re-recording mixer / sound designer / supervising sound editor," favored "organic as opposed to synthesized sounds" and spent much time during the shoot "in Jordan recording Muslim calls to prayer and car horns."[3] Ottosson shared his effects sound files with the composers Marco Beltrami and Buck Sanders. Beltrami studied composition at the University of Southern California under film composer Jerry Goldsmith, and Sanders specializes in signal processing: the pair combine the competencies of a conventional Hollywood film music composer and a computer-oriented producer. Erhu player Karen Han contributed to both effects and score. The erhu is a two-string, fiddle-like, melodic instrument played across East and Southeast Asia. Ottosson described Han improvising while "looking at the movie . . . then I'll take that and modulate it, tweak it, and it really doesn't sound like an erhu anymore, but the emotion is there."[4] Han called these contributions not "musical things" but just "sound for [Ottosson's] sound design" and compared the erhu "to a human voice used as an instrument and it cries when I play."[5] Han is also the primary sonic element in the heavily processed lyrical theme, almost the only recognizably "musical" content in Beltrami and Sanders's score. Director Kathryn Bigelow initially did not want a theme for the score, but the picture editors convinced her to include one after using music from Beltrami's score for the 2007 Western *3:10 to Yuma* as a temp track (temporary music used during the editing process). Beltrami and Sanders delivered their score for the film in a state where it could be completely remixed. (Most composers deliver scores mixed into several "stems"—for instance strings, brass, percussion—limiting the extent to which individual elements of the mix might be adjusted. Score albums often present a mix where the stems are mixed noticeably differently from the score in the film; see the discussion of *United 93* in chapter 9.) The score album for *The Hurt Locker*, produced by Beltrami and Sanders, is best understood as a remix by the composers of elements they created for, and let others manipulate in, the film.

Apocalypse Now and The Hurt Locker—the former psychedelic, the latter quasi-documentary—promise immersive experiences grounded in a breaking down of the borders between sound effects and music, all in the service of representing war as an extreme, total sensory experience, with sound central to the effect.

Approaching the soundtrack in integrated fashion as a combination of dialogue, effects, and music allows for a film whole to be understood as a sonic whole, as a species of time-based sonic art, as having the status of music. This analytical embrace of all types of sound puts diegetic and nondiegetic distinctions in the music track into perspective, recognizing how the soundtrack is, above all, a construct that lends selective sonic life to the image track. Integrated soundtrack analysis also inserts silence into the mix. Paul Théberge has usefully broken down several kinds of film silence.[6] His term *musical silence* offers a reminder that the absence of music is a musical choice. For example, the D-Day sequence in *Saving Private Ryan* unfolds in musical silence. When the battle is over, music comes in as dialogue and effects fade out. Théberge's reciprocal term *diegetic silence* describes a mix where dialogue and effects are muted in favor of nondiegetic music (or background scoring) only or silence in all sonic domains.

Indeed, diegetic silence proves an essential category for PCF soundtracks. To remain with *Saving Private Ryan*, the opening and closing sequences with old Ryan at the Normandy cemetery unfold largely without effects. The opening includes but one word of dialogue ("Dad"). The film's close might have pleased critics more if the dialogue between Ryan and his wife about whether or not he was a good man had been omitted, leaving only sympathetic looks and Ryan's final salute to Miller's grave marker to the strains of Williams's score. *Courage Under Fire* and *Act of Valor* (2012) close with similar visits to cemeteries that unfold in diegetic silence with music and voice-overs spoken by the dead. This choice leaves the conclusion of these films more open to interpretation. Fostering ambiguity in the PCF, diegetic silence—often filled with music—proves an important soundtrack strategy.

NARRATIVE SHAPES

Film soundtracks are driven by film narratives, which are, in turn, fundamentally shaped by genre. In the musical, characters sing—and so music and singing prove especially important as an organizing feature of musical film soundtracks. In the laconic genre of the Western, climactic shoot-outs often rely on near-total silence shattered by brief outbursts of gunfire. Western soundtracks are, at times, driven by sound effects. War film narratives include scenes of combat, which tend to make war film soundtracks especially loud—noisy with the sustained sounds of guns and explosions near and far. When war arrives in a dialogue-driven genre such as melodrama, noisy effects intrude on the soundtrack, enhancing the experience of the onset of war as a disruption of everyday life. For example, *From Here to Eternity* (1953) unfolds as a rather quiet drama until the story arrives at the morning of December 7, 1941, when Japanese planes bomb Pearl Harbor. Then, for a stretch of minutes, the soundtrack is dominated as nowhere else in the film by deafeningly loud sound effects. The sequence begins in a kitchen, where pots and dishes crash to the floor, transitioning into the even noisier sounds of planes and machine guns. The battle plays out with minimal dialogue—typical of most extended combat scenes—and receives no musical enhancement. For these minutes, *From Here to Eternity* sounds like a war picture.

Understanding how sound (generally) and music (specifically) function in PCFs begins with an examination of narrative types in the subgenre, which pulls by its very definition against many of the conventions of Hollywood film. PCF scores are not usually keyed thematically to individual characters (though a prominent few described later in this chapter are) and never work to underline romance plots. Instead, music tends to serve structural purposes: marking the enemy's frequently unseen presence, spreading a sonic field of danger, putting a popular-music background behind the conversations of like-minded fellows, or opening up a stretch of time wherein ceremonial acts, such as funerals, unfold. A rather limited set of narrative structures have served the stories the subgenre tells. A survey of narrative shapes in the subgenre follows as preparation for the initial consideration of PCF scores that closes the chapter.

Some PCFs adopt a two-part structure, dividing the film into two large sections that are sometimes marked by different approaches to musical scoring. One version of the two-part structure follows a group of men as they move through boot camp then into the battle zone, demonstrating along the way how military manhood is made and the bond of brotherhood is built. Such stories often begin stateside at a military base, as in *The Boys in Company C* (1978), *Full Metal Jacket* (1987), *We Were Soldiers* (2002), *Windtalkers* (2002), and *Jarhead* (2005). In *Hamburger Hill* (1987), new arrivals in Vietnam are rushed through a behind-the-lines initiation before being tossed into the fight. A second form of two-part narrative moves from preparation to execution of a battle plan. Typically professional soldiers (and the movie audience) are first briefed on, then execute, a battle plan, which sometimes goes tragically awry (*Black Hawk Down* [2001], *Lone Survivor* [2013]). *Letters from Iwo Jima* (2006) begins by following the Japanese commander as he prepares the battlefield for the imminent US invasion. The abrupt arrival one morning of the US fleet marks the start of the film's second part. *United 93* (2006) presents a special two-part case. The tense early scenes of the film center on the hijackers preparing for the mission, boarding the plane, and waiting to take control of the aircraft. Once the hijacking of United 93 commences, the nature of the film—and of the score—changes appreciably.

The relative proportions of the two parts prove a crucial formal aspect in all of the above films. Some are about evenly balanced (*Full Metal Jacket*). Others are less so, shifting to battle action early on (*Hamburger Hill, Windtalkers, Black Hawk Down*). Often the transition between parts is articulated musically. In *Hamburger Hill, Black Hawk Down, We Were Soldiers*, and *Lone Survivor*, a helicopter ride featuring prominent music marks the move to the battlefield. (These transitions are compared in chapter 7.) *Jarhead* toys with the training to combat structure, since combat per se, for the film's characters, never quite occurs.

Two Vietnam films—*The Deer Hunter* (1978) and *Born on the Fourth of July* (1989)—expand the two-part model into a three-part structure. Both films spend their first parts concentrating on soldiers' home lives rather than their military preparation. After crucial battle sequences set in country, the films return home to explore the effect of the war on those who survive. Combat experience forms the centerpiece of both dramas.

Zero Dark Thirty (2012) uses a three-part structure fitted to the nature of combat in the GWOT. The first section dwells on the use of torture: for thirty minutes the film moves through CIA black sites where detainees are subjected to "enhanced interrogation." When the CIA investigators happen upon a name that will, in the end, lead them to Osama bin Laden's compound, the film shifts in pace markedly—and the musical score changes as well. An eighty-minute central section builds across a series of successful terrorist attacks as the main character, Maya, concludes that she has indeed found bin Laden's hiding place. The third section, the film's final fifty minutes, begins with the introduction of the Navy SEALs who execute the bin Laden raid. The SEALs and their Stealth helicopters make a dramatic entrance heralded by the opening of huge hangar doors. All three sections of *Zero Dark Thirty* are legible as combat in a GWOT context. The transitions between sections occur gradually, with changes of narrative mode often marked first by the introduction of new musical themes or textures.

A second, larger narrative shape follows an alternating pattern of action and reflection, formally expressing the diastolic rhythm of soldiers' lives. As the 1946 book *Soldiers' Album* put it: "Combat was the exciting and the terrifying part of war, but it wasn't all combat. Mostly it was walking or driving or waiting."[7] A similar, more routine movement inside and outside the wire was described by Iraq War explosive ordnance demolition (EOD) tech Brian Castner: "And so every day the pendulum swung. Danger to safety. Safety to danger."[8] In his 1959 book *The Warriors*, World War II veteran J. Glenn Gray generalizes, "The alternation of dullness and excitement in their extreme degrees separates war from peace sharply."[9] Typically these two narrative modes—danger and safety, dullness and excitement—are represented in the PCF by alternating scenes of mostly fighting and mostly talking, on the soundtrack by alternating prominence given to sound effects and dialogue. Music can be incorporated in scenes of either action or reflection, depending on the larger aesthetic strategies informing the score.

Several signal films in the subgenre follow this back-and-forth narrative structure. *Platoon* (1986) affords a classically balanced example. Four times the platoon goes on patrol "outside the wire" in the jungle: on each patrol they encounter the enemy in a different combat mise-en-scène (isolated nighttime ambush; daytime encounter with tunnels and booby traps that ends in the platoon terrorizing a village; a rainy daytime running battle through a muddy forest; a climactic, full-on night attack with flares creating eerie shadows). Between these battles, the platoon goes back to base camp for three stretches of rest, conversation, and—in one instance—music and dancing. *The Hurt Locker* similarly unfolds as a series of six action scenarios, each involving a different sort of bomb or enemy engagement, alternating with behind-the-wire reflection and conversation among the three primary characters. Amy Taubin called the film "a structuralist war movie—it could be titled 'Seven Instances of Dismantling an Improvised Explosive Device.'"[10] Chris Kyle in *American Sniper* (2014) journeys to Iraq for four tours of duty, each marked on-screen by a title. A viewer familiar with Kyle's best-selling book can anticipate that the fourth tour is the last. The reflection sequences in *American Sniper* take place stateside and often begin with disconcerting abruptness, suddenly dropping the viewer back into the world of home, where conflict of a domestic sort awaits. The epigraph and poster tagline for *Act of Valor* reads, "Based on real acts of valor," signaling both the authenticity of the combat scenarios and the film's compartmentalized mission-oriented plot, which is reinforced by constant and exotic shifts in location around the globe—marked, as in many action films, by animated graphics, including maps.

Some action-reflection narratives work along a strongly driven narrative line; others are more episodic. *Go Tell the Spartans* (1978) combines alternation between battle and reflection with a well-developed through line, moving dialectically toward a demonstration of the misguided US strategy of fortifying jungle outposts. *Three Kings's* (1999) heist plot and *The Kingdom's* (2007) procedural trajectory move with equally strong intent: the former along an unexpected path, the latter according to a "more conventional format or architecture . . . where you kind of hunt down the bomb-maker" (to quote Bigelow's description of what she wanted to avoid in *The Hurt Locker*).[11] *In the Valley of Elah* (2007), also a procedural that arrives at the expected solution to the case, punctuates the generic rhythm with fragments of recovered combat footage: glimpses of Iraq that contribute nothing to solving the mystery but instead disturb the narrative and its main character with troubling context that threatens to expand the film's implications beyond a single military family's trauma to the broadest national level. *Apocalypse Now* is both picaresque and strongly motivated: a series of varied encounters, not all combat related, are separated by movement upriver in a patrol boat (PBR) marked by talk among the men and ruminative voice-overs by Willard, the soldier sent to terminate Colonel Kurtz—the mission of the film—as he goes through Kurtz's dossier. *Saving Private Ryan* and *The Thin Red Line* (1998) alike follow the pattern of combat action and reflection. In the former, the men either move forward in the search for Ryan—their journey is all on foot—or stop to engage the enemy or rest. The deaths of two among their group mark stages along the way. The mechanics of the plot are expressed in a repeated action—the passing on of a letter home—intended to increase in poignancy with each repetition. Action and reflection in the less-directed narrative of *The Thin Red Line* are marked by active engagement with the enemy and periods of rest on transport ships or off the line. Movement to, across, and off the island of Guadalcanal provides a loose spatial trajectory for the plot, such as it is. In both films, the natural end point of the narrative comes with the death of the central figure: Miller in *Saving Private Ryan*, Witt in *The Thin Red Line*. *Red Tails* (2012) follows this pattern as well—climaxing with the death of a main character, if along a messier track due to multiple, underdeveloped plot strands.

The spectacle of combat, and the implied promise to the viewer that battle action will drive the narrative, is central to nearly all PCFs. Based closely on an actual event, *Casualties of War* (1989) nonetheless toggles back and forth between action and reflection—with the caveat that among the violent events are the gang rape of a kidnapped Vietnamese woman, and an attempt to frag a soldier who has informed on the rape and subsequent murder of the woman. The film opens with a thrilling action sequence unrelated to the central plot that assures the viewer that *Casualties of War* is, indeed, a combat picture. The fighting-talking structure present in so many of these films secures their generic identity as action films. Audiences expect action films to unfold as a series of action sequences—the equivalents of musical numbers in a musical. Scenes of talking are expected, too, but only as a respite from the set pieces that are the crucial stuff of the action genre. Even a philosophical war film like *The Thin Red Line* meets the generic expectation that action and dialogue will alternate in reasonable proportion. (With only two combat scenes at either end of the film and a welter of plots and characters, *Miracle at St. Anna* [2008] is difficult to categorize as a combat film in terms of its narrative structure.)

Films with two-part narratives sometimes fall into an alternating action-reflection pattern once they arrive at the battlefield. Shifts between combat action and dialogue scenes in the second half of *Full Metal Jacket* are echoed by similar, not always fully aligned, shifts between unscored scenes and scenes scored with popular 1960s records. *Hamburger Hill* periodically marks the progression of the ten-day battle with titles, such as "15 May." The structure works well to express what William Pelfrey called the "endless circular ritual" of the war in Vietnam.[12] In its long second part, *Windtalkers* follows a highly formalized alternation between fierce, exceedingly violent, epic and close-range fighting and dialogue scenes turning on the cultural divide between the Navajo code talkers and the white Marines. Each of the combat scenes climaxes with a very large explosion, reinforced loudly in the musical score, effectively signaling to the listener that the sequence is over. A verbal exclamation on the order of "cool" by one of the Marines in the story further underlines these moments in the dialogue domain. Other recurring transition markers in the film include voice-overs: radio chatter among unidentified code talkers in Navajo, which is translated in subtitles, and gentle, caring letters, which go unopened by the main white character, read in the voice of a nurse briefly seen during the training portion of the narrative. These overlapping devices lend *Windtalkers* a kind of formality, making perhaps overly manifest structural patterns usually used more subtly.

In some twenty-first-century films, the action is so continuous that scenes of reflective dialogue are reduced almost to nothing. *Black Hawk Down* is the innovator here, and almost every reviewer commented on the absence of downtime, previously a regular part of war film pacing. The combat film historian Jeanine Basinger called *Black Hawk Down* "like nothing I have seen before," noting how virtually the entire run time "distills a pure combat experience, disorienting, chaotic and bloody."[13] *Green Zone* (2010) hardly takes a breath as it bounces between violent confrontations in the streets of Baghdad and verbal confrontations in the confines of the Green Zone. The final verbal confrontations turn physical—marking a point of complete breakdown in US efforts in post-invasion Iraq. This level of narrative forward motion creates the conditions for both films' almost continuous beat-driven scores (see chapter 9).

The three combat television series included among PCFs adapt the above two narrative types to their episodic nature and much longer time span. *Band of Brothers's* (2001) first episode functions as preparation for the D-Day invasion, which is depicted in episodes 2 and 3. Every subsequent episode, except for the ninth, features some sort of combat action alternating with conversation among the men. Episode 5 offers a self-reflexive version of action-reflection. A series of combat actions are presented as the memories of Captain (later Major) Dick Winters, who in the present tense of the episode is typing a detailed after-action report. Sound plays a formal role, with cuts between past and present matched by way of sound effects: gunshots on the battlefield become the sound of Winters's typewriter. The process of turning combat experience into fixed narrative—whether a formal report or, by analogy, a film or television series—is expressed by one sound representing both action and the act of reflecting on action. Battle action occurs with less consistency in *The Pacific* (2010). Indeed, episode 3 includes no combat at all and centers on a romance between a Marine and an Australian girl right out of pre-Vietnam military melodramas such as *Battle Cry* (1955). *The Pacific* is also narratively burdened by its braiding together of the stories of three real Marines: Robert Leckie and Eugene Sledge, who both wrote combat memoirs that dwell on the dark side of the soldier's experience, and John Basilone, a winner of the Medal of Honor who lived an unabashedly heroic life and died on the beach at Iwo Jima. The narrative style of the series shifts as the focus passes between these three men. After its pre-invasion first episode, *Generation Kill* (2008) unfolds as an oscillation between action and reflection, where the former often feels to the men like frustrating inaction punctuated by sudden life-or-death encounters (such as manning a checkpoint). Indeed, one of the achievements of the series is its depiction of the boredom of the battlefield—a risky choice if the goal is to hold the viewer.

The vast majority of PCFs immerse the viewer in a chronological unfolding of events. Only two play substantially with time: *Courage Under Fire* and *Flags of Our Fathers* (2006) both search for the truth of specific battlefield events by jumping backward and forward in time. *Courage Under Fire* plays with perception and deception by presenting several different versions of the same event in a *Rashomon*-style, flashback-heavy narrative. The need to relive combat experiences to escape their debilitating influence forms a pillar of the film's secondary plot: a sound recording of a battle is played to assure the main character that he acted correctly while "under fire" despite having made a terrible mistake. More radically, *Flags of Our Fathers* abruptly cuts between several levels of past and present: the "now" of a son discovering the "then" of his father; the lingering traumatic battle memories of soldiers reliving the deaths of their buddies; the immediate trauma of battle itself. Indeed, the narrative is so splintered that it becomes difficult and unnecessary while viewing to definitively assert which scenes are flashbacks and which are the "present."[14] Generically unusual approaches to narrative time lend both *Courage Under Fire* and *Flags of Our Fathers* a distant quality. The viewer is enjoined to think about combat and its effects on the soldier without necessarily being immersed in a combat narrative, although the satisfaction of discovering what *really* happened drives both films.

SCORE TYPES

Having traced the larger patterns of PCF narratives, the overall shape and content of PCF scores can be assessed. Music proves tremendously important formally and expressively in almost all of these films, one reason they welcome the designation as hymns. What work does music accomplish in the PCF that makes its prominent use so pervasive? Three basic questions will guide the following survey of the subgenre:

1. How much of a film's run time is scored?
2. How is the score distributed across the film?
3. What kinds of music are used when and where?

Answering these questions in a general way here provides a foundation for the consideration of soundtracks and scores in greater detail in later chapters. The felt stylistic similarities or contrasts between films can often be understood at the level of large-scale form as expressed in the use of music.

The first and second questions concern form; the third, content. Taking up content first, film scores include all the music in a film, whether composed especially for the film (original scoring) or borrowed from elsewhere (compiled or interpolated cues). Borrowed music can come from anywhere: from classical music, which carries a kind of historical transparency for most listeners, or popular music, which, as Anahid Kassabian has argued, carries "the immediate threat of history," since audience members potentially "bring external associations with the songs into their engagements with the film."[15] External associations with popular music, often generational by nature, are crucial to some but not all PCFs. Close connections between the popular music of the 1960s and the Vietnam War, combined with the prominent use of nostalgic popular music in film scores to activate particular generations of moviegoers, proves foundational to the 1980s PCF cycle. The World War II and post-9/11 cycles do not—indeed, cannot—deploy popular music in this way. Comparing the borrowed musical content of PCF cycles reveals how popular music history has shaped filmmakers' ability to leverage music in the subgenre.

The form-oriented questions of extent and distribution can be understood by way of four general, at times overlapping, categories:

- films with minimal music
- films that alternate between scored and unscored scenes
- music-laden films, with music noticeably present a lot of the time
- films with almost continuous music

Below I sort all thirty-five members of the subgenre into these categories, providing a unified view of the films and their use of music. The goal here is to understand music as a species of film sound at the broadest, most comparative formal level. Subsequent chapters treat select groups of films along more narrowly drawn topics relating to recurring sonic tropes (such as soldiers' singing or the sound of helicopters), specific sound and music techniques (such as voice-over or beat-driven music), and specific musical registers (such as the elegiac).

Films with Minimal Music *(The Boys in Company C, Hamburger Hill, 84 Charlie MoPic, Generation Kill, American Sniper)*

The few PCFs that use minimal music provide a useful contrast to the majority of the members of the subgenre.

84 Charlie MoPic (1989) offers an extreme. This single-camera mock documentary following a patrol through the jungles of Vietnam includes only the image maker—the title character, MoPic, an Army filmmaker seeking "lessons learned"—and almost completely elides the necessary collection of sound. MoPic mentions a "slate," which would mark a synchronization point between image and sound recordings, and a microphone is briefly seen at the start, but no soundman travels with the soldiers in the film. The setup with MoPic alone collecting sound and image more accurately reflects the camcorder technology of the 1980s, when the film was made, than it does the separate sound and image capture of television or film crews during the Vietnam War. All other instances of a film crew in a Vietnam film—see *Apocalypse Now, Hamburger Hill,* and *Full Metal Jacket*—include a soundman with recorder alongside a cameraman with camera. There is very little music in *84 Charlie MoPic*: gentle acoustic guitar backs the film's dedication to four specific Vietnam companies before the narrative begins; the men pick up the Armed Forces Vietnam Network briefly on their field radio; and there's music during the end titles. *84 Charlie MoPic* includes virtually all the narrative tropes of the combat genre *without* musical support. It is, in that sense, a control of sorts. What other films choose to score—for example, the helicopter ride into the jungle—*84 Charlie MoPic* leaves in a state of musical silence.

The Boys in Company C and *Generation Kill* follow *84 Charlie MoPic* in using no scoring and very limited source music (a strategy that finds a pre-Vietnam precursor in *Battleground* [1949]). *The Boys in Company C* opens with "The Marines' Hymn" sounding behind the film's main titles. After this, all of the music is made by the men of Company C or other diegetic sources. The use of music is even more minimal in *Generation Kill*. Only one piece of recorded music is featured during the entire 489 minutes of the series: Johnny Cash's "The Man Comes Around" plays as the sound-track for a YouTube–style video compilation assembled by one of the men and watched by all at the end of the final episode.[16] Other than this, virtually the only music heard is snatches of pop music sung by the men themselves. The opening and closing titles use only radio chatter and overheard conversations, which, as described in chapter 5, act as music. In this *Generation Kill* differs greatly from *Band of Brothers* and *The Pacific*, which both have lush opening and closing music framing every episode and abundant narrative scoring throughout. *The Boys in Company C, 84 Charlie MoPic*, and *Generation Kill* limit musical content with great discipline, an approach not typical of the subgenre.

Hamburger Hill does use some music. Philip Glass's score—a single cue heard at the top and bottom of the film and twice briefly late in the battle—was discussed in the introduction. Ten pop-music cues—eight familiar American records and two Vietnamese tracks—are heard briefly as fragmentary source music during the relatively short preparation phase of the narrative. None take on much prominence in the mix, and all work mostly to enhance the viewer's sense of time and place. However, one interpolated cue in *Hamburger Hill* carries great importance. At the hinge in the narrative, on the helicopter ride into battle, the Animals' "We Gotta Get Outta This Place" is heard almost in its entirety. Director John Irvin chose the song as a riposte to Francis Ford Coppola's use of Richard Wagner's "Ride of the Valkyries" during the helicopter battle in *Apocalypse Now*.[17] These and other musicalized helicopter sequences are compared at length in chapter 7. Here, it's worth noting that *Hamburger Hill*, a combat film with very little music, shifts toward a highly musical mode at a crucial narrative juncture and in response to an earlier film that constructed American fighting men and their eagerness to fight by way of a strong musical choice. The intertextual argument between these films is explicitly musical.

American Sniper has no credited composer. Director Clint Eastwood composed "Taya's Theme"—a one-finger piano melody heard three times far back in the mix, always in association with Kyle and his wife Taya's sexual chemistry. Joseph DeBeasi, a Hollywood-based music editor credited with "additional music," likely contributed the film's two isolated, beat-driven cues—taken together less than two minutes of music. And yet the soundtrack of *American Sniper* is alive with many very short—seconds-long—bumps and pulses of metallic sound, often heard on hard cuts between scenes (the image track has but one fade-out, at narrative's close). Humming backgrounds also come in and, a few times, a low, heartbeat-like beat. These sonic elements are small, micro enhancements of a film that otherwise clings to a kind of documentary-style, camera-oriented audio perspective, almost entirely resisting music as a softening agent—until the fulsome music for its end titles.

Alternating Scored/Unscored *(Go Tell the Spartans, The Deer Hunter, Platoon, Courage Under Fire, Saving Private Ryan, Band of Brothers, United 93, In the Valley of Elah, Redacted, The Hurt Locker, The Pacific, Red Tails, Zero Dark Thirty, Lone Survivor)*

Many combat films alternate between scored and unscored scenes.

The press kit for *Saving Private Ryan* articulates this strategy in explicit terms: "Williams and Spielberg mutually decided which scenes should have music. To accompany the journey of Miller and his squad, they chose to have the music flow out in long sequences, followed by scenes with no music at all."[18] Alternation of scored and unscored scenes yields the following pattern in *Saving Private Ryan* measured in minutes with music (in bold) and without music:

$$4 / \underline{\underline{23}} / 9 / 5 / 1 / 37 / 5 / 8 / 4 / 5 / 4 / 9 / 5 / .5 / [8] / \underline{\underline{27}} / 8 / 6$$

The double-underlined unscored portions of twenty-three and twenty-seven minutes at either end of the film are the major battles: D-Day and the defense of the bridge at Ramelle. The final, six-minute cue, "Hymn to the Fallen," plays during the end titles. All the music in the film is scoring except for the bracketed eight-minute cue, which uses three records by the French singer Édith Piaf played as source music on a phonograph discovered in the ruins of Ramelle. The conversations during this stretch are effectively timed to the length of these records. Only a few seconds of Williams's score accompanies battle action, which instead unfolds almost entirely without musical support. Indeed, Spielberg articulated this goal for the score in his liner notes for the score album: "Restraint was John Williams's primary objective. He did not want to sentimentalize or create emotion from what already existed in raw form. *Saving Private Ryan* is furious and relentless, as are all wars, but where there is music, it is exactly where John Williams intends for us the chance to breathe and remember."[19] The alternating structure of the narrative—moving back and forth between action and reflection—is echoed in large part by the score. Spielberg's two combat television series largely follow this pattern within their much longer time spans and more complicated narratives, with more music in *The Pacific*, including some scored combat action, than in the comparatively more restrained *Band of Brothers*.

Alternating scored and unscored scenes work in much the same way in *Platoon*'s score, which mixes original scoring and compiled classical and popular music. The background score has only seventeen cues: eight use Samuel Barber's *Adagio for Strings*, and three use the film's nominal composer Georges Delerue's attempt to copy the Barber (which began as *Platoon*'s temp score and was privileged by Stone over Delerue's effort). The formal alternation of scored and unscored often takes an ABA form in the film. For example, at the burning of the village the Barber fades in as Zippo lighters set thatched roofs aflame, re-creating an iconic image of the war. The music plays on as the village is destroyed. It cuts off at a dramatic rest (or silence) in the music just after four high fortissimo chords and just before Chris stops some men of the platoon from raping a young girl. This short encounter, with important dialogue, is left unscored. The Barber then restarts—just about where it left off in the piece—and plays to the end of the village sequence before fading to silence at the end of a phrase in the music on a transition to a base camp sequence. The entire sequence is selectively and sensitively cut by Stone to the shape of Barber's music. Similar ABA structures occur on three other occasions in *Platoon* (each discussed in subsequent chapters).

Films that alternate between original scoring and musical silence include *Go Tell the Spartans, Courage Under Fire, In the Valley of Elah, United 93, The Hurt Locker, Red Tails, Zero Dark Thirty,* and *Lone Survivor*. None of these films have significant compiled popular music cues. In all eight, an original score acts in a restrained manner, adding expressive content that generally does not call attention to itself and seems to be supplementing rather than driving the unfolding of the film. (Exceptions to this general characterization of these films, in particular cues from *The Hurt Locker, Red Tails,* and *Lone Survivor,* are discussed in later chapters.)

Two further films—*The Deer Hunter* and *Redacted* (2007)—fall into the alternating scored and unscored category due to their lack of a significant original score, inclusion of a variety of musics, and targeted use of musical moments to make a local impact. Neither has a larger strategy for musical meaning across the whole, yet both have powerful musical endings.

The Deer Hunter's main and end titles music is a borrowed melody: "Cavatina" by Stanley Myers. Much of the film's long opening sequence—a Russian Orthodox wedding and reception in a Pennsylvania steel town—unfolds to music performed within the scene, whether by the church choir or the wedding band, lending the whole a quasi-documentary quality. On arrival in Vietnam, the film's battle and prison camp scenes—including the games of Russian roulette—are unscored. Many of the original score cues, also by Stanley Myers, are very short and none cohere thematically or draw the viewer into a larger net of musical meaning. And yet *The Deer Hunter* includes three especially resonant musical moments—all diegetic, all set in a bar owned by a friend of the trio of main characters. Early in the film, the Four Seasons' "Can't Take My Eyes Off You" blasting from a jukebox prompts an explosive all-male sing-along.[20] A Chopin nocturne played by the bar's owner, John, brings the film's first part to a close: the sound of a helicopter fades in while the image of men listening in the bar lingers on-screen, initially activating the central Vietnam section on the soundtrack. Most famously, the singing of "God Bless America" by the surviving characters in the final scene was mentioned in most every review (see chapter 4). Like its score album, *The Deer Hunter* is a hodgepodge of musics—many put to great but targeted effect—that hold together only by virtue of their being used in the same film.

The narrative of *Redacted* accumulates slowly by way of more than ten kinds of documentary footage: a soldier's video war diary, a professionally made French documentary (which uses the same stately sarabande by George Frideric Handel over and over), clips from Iraqi and European television news, video from American and jihadi websites, video documentation of psychological exams and court depositions, a surveillance camera (which, unusually, also has sound capabilities), video chat between a soldier and his father, and home video of a welcome-home party. Music is heard as appropriate in these sources, and there isn't much music in the film. However, the film's final (and only implied score) cue—an orchestral version of an Italian opera aria by Giacomo Puccini—gathers the emotional weight of the entire narrative into a musical release (see chapter 11).

While neither *The Deer Hunter* nor *Redacted* use music in a consistent manner, both end with intensely musical moments that serve as key interpretive sites for viewers of these films.

Music-Laden Films *(Full Metal Jacket, Casualties of War, Born on the Fourth of July, Three Kings, We Were Soldiers, Windtalkers, Jarhead, Flags of Our Fathers, Letters from Iwo Jima, Miracle at St. Anna)*

Another set of films, many among the most ambitious in the genre, use music prominently throughout. The viewer's awareness and enjoyment of the music proves central to the experience of these films, which might be called lyrical, or even musical, combat films. They are sorted here by content. Three (*Full Metal Jacket, Three Kings, Jarhead*) combine compiled popular music cues deployed for maximum effect in the mix with an original score that also does important structural and expressive work, creating what the film music scholar Jennifer Psujek terms *composite scores*.[21] A second trio (*Casualties of War, Born on the Fourth of July, We Were Soldiers*) uses expansive, musically grand, original orchestral scores to draw out larger narrative themes along long-established Hollywood lines. Eastwood's Iwo Jima films make effective use of rather modest musical materials. (The extensive original scores for *Windtalkers* and *Miracle at St. Anna* fall outside larger trends in individual ways not germane to this study.)[22]

Full Metal Jacket is uncommonly full of loud music. The film begins with the slide guitar intro to the 1965 country song "Hello Vietnam," which plays loudly over slightly brutal images of Marine recruits having their heads shaved, a roll call of the cast at the start rather than the conclusion of the film. The razor shaving off all that hair looks ferociously powerful, yet we never hear its buzz, which is probably more like a roar. (A similar haircutting scene in *The Boys in Company C* suppresses image in favor of only the fearsome sound of military barbers making quick work off camera.) The first half of *Full Metal Jacket*—set on Parris Island, where Marine recruits are turned into Marines—unfolds in three related sonic modes: shouted, profane, yet poetic harangues by Gunnery Sergeant Hartman; Jody chants (or call-and-response cadences that keep men moving together in time) sung by Hartman and the Marines, which put the sergeant's foul language into rhyme and meter; and an instrumental mix of drums and low brass by the composer Vivian Kubrick (billed as Abigail Mead), which draws musically on the Jody chants. The three modes trace out a spectrum from heightened dialogue to music and cast a highly ordered structure over the training portion of the film. The only other music in the film's first half is "The Marine's Hymn," played at the graduation ceremony as a climax (of sorts), and a haunting musical texture used in relation to Private Pyle, the tormented recruit who murders Hartman before killing himself at the climax and close of the half. The second part of *Full Metal Jacket*, set in Vietnam, alternates between unscored scenes and scenes played with very loud compiled cues drawn from Vietnam-era popular music in the background score. The pop tunes play mostly in their entirety. This segmented approach—bracketing off scenes by way of the presence or absence of music—prevails until the very end of the film, when the haunting music heard at Pyle's violent end returns for the point-blank killing of a female Vietnamese sniper. (This music of dread is analyzed in chapter 8.) *Full Metal Jacket*'s final credits roll over a complete playing of the Rolling Stones' "Paint It Black." On many levels and in many ways, Stanley Kubrick's film builds film form on musical form.

Following the example of *Full Metal Jacket*, *Three Kings* and *Jarhead* both use a combination of original scoring and prominent pop records. However, in these films the two kinds of music are not separated into discrete narrative sections but instead alternate freely. And unlike *Full Metal Jacket*, which denies its characters any evident pleasure in pop music, *Three Kings* and *Jarhead* both use diegetic music to show military men celebrating with each other in memorable vignettes of group joy or frustration. Music-laden films provide ample opportunities for this approach to characterizing the fighting man. (Soldiers sharing music forms the central topic of chapter 4.)

Casualties of War, *Born on the Fourth of July*, and *We Were Soldiers* all have grand orchestral scores—by Ennio Morricone, John Williams, and Nick Glennie-Smith, respectively—that are put primarily in the service of characterization. Most cues in these films work to evoke empathy for primary characters. Scoring also plays during battle scenes in all three, a mark of the importance of music in these films, as PCFs tend to leave the combat soundscape to effects and minimal dialogue (see chapter 6). While these films include a few popular music cues—*Born on the Fourth of July* has the most—almost all are source music working more or less as musical wallpaper. Morricone, Williams, and Glennie-Smith's scores alike rely on strong melodic themes that are developed or simply reiterated, hearkening back to classical Hollywood practices. When attended to closely, these lengthy and substantial scores add a layer of storytelling and commentary to each film.

Morricone's music for *Casualties of War* develops two contrasting themes: an elegiac melody used sparingly to score American soldiers, and a cluster of related modal themes used extensively in relation to Oanh, the Vietnamese civilian victim of rape and murder at the center of the story. Morricone effectively amplifies Oanh's plight throughout the film: she is the main character as far as the score is concerned, and the music for the end titles confirms this (see chapter 11). Music-laden films allowing ample space to original scoring give composers extra opportunities to shape film meaning.

Oliver Stone imagined the score for *Born on the Fourth of July* in his earliest script, where he repeatedly spots an original "MUSIC THEME" described in evocative and contradictory terms: "Spring. Youth. Suggesting an onrushing force"; "in one sense melodic, stirring, in another rushing, pulling, sucking Ronnie from his youth, pushing him on into manhood"; "pressing, destiny-ridden."[23] Late in production Columbia added to the budget, enabling Stone to engage John Williams, Hollywood's preeminent composer at the time.[24] Williams's score uses several themes, which flow easily into and out of one another. One, however, takes on special prominence: heard in substantially the same form seventeen times, this rounded melody, with alternate minor and major mode versions, features a pair of equal-length phrases, easily repeated or invoked, in the cellos and basses. Heard above a tonic pedal, the predominantly stepwise melody alternately circles around the dominant and tonic pitches—just so much frustrated motion going nowhere, an expression of reserved and contained grief that cannot be escaped, a futile dwelling on loss. Williams uses this melody across the film's long narrative, which remains throughout centered on Ron Kovic, who appears in every scene. The theme casts a shadow over Kovic's 1950s childhood, from his short voice-over description of playing war games as a boy to the young Kovic watching crippled veterans roll by in a Fourth of July parade. Later he himself will be a veteran in a wheelchair. The theme plays over an extreme close-up of Kovic's deadened eyes as he returns to the battlefield after accidentally killing a young American soldier and being rebuffed by his commanding officer, who refuses to hear his confession. It sounds just after his initial wounding in the foot, and as a priest administers him last rites in the chaos of a field hospital. The theme stays with Kovic on his return to the United States, marking his painful passage to acceptance of his physical condition and embrace of the antiwar movement. A poignant use comes on Kovic's return home to his family after a horrific stay in a veterans' hospital (a sequence scored entirely with upbeat, soulful pop tunes that almost make the physical humiliations of the hospital watchable). Williams's most-used theme accompanies a point-of-view shot from the perspective of Kovic's mother as she walks out to greet him in their front yard, publicly embracing her wheelchair-bound son for the first time. Mother and son assure each other that each "looks great"; the score registers very different feelings, which boil over in an unscored shouting match several scenes later. Williams's score also sounds a more uplifting note by way of several themes associated with home and healing. Like the mournful theme described above, these secondary themes rely on a strings-only sound—with some solo oboe and trumpet—and often utilize overlapping contrapuntal voices. Stone understood from the start that his complex central character would need unified yet flexible musical support, a score with the emotional sweep of classical film music that would tie the sprawling story together and characterize Kovic's inner life with all its contradictions: the passionate belief of his youth, his disillusionment after Vietnam, his recovery of a sense of mission at the close. Working on a broader scale and with much greater flexibility—and heard more consistently across the film—than the interpolations of Barber's *Adagio for Strings* in *Platoon*, Williams's score, enabled by major studio financing, sets a sustained tone, interpreting every image over which it plays as a moment to reflect on loss, serving an elegiac function that underpins Stone's cinematic eulogy for America in the 1950s and 1960s. However, in his soaring music for the end titles, Williams endeavors to break through to some measure of transcendence, setting aside the oft-repeated mournful theme and dwelling only on the themes of home and healing, answering Stone's wish, put into words in his final script draft, for a "MUSIC THEME now . . . of tragedy overcome. Of life renewed."[25] Here again a composer's "last word" directs the listening movie watcher to a particular reading of the film just seen.

We Were Soldiers has a lot of music. The opening titles cue suggests that the film and its score will offer generic action-movie thrills. Instead, the score explores the elegiac register central to PCFs from *Platoon* onward (discussed in detail in chapter 10). Glennie-Smith's main theme is first heard very early in the film: a short, lyrical cello melody anoints a young officer named Jack Geoghegan, played by Chris Klein, as he prays in a hospital chapel after the birth of his daughter. This theme, planted near the film's start, returns about an hour of screen time later when news of the first combat deaths reaches the wives back in the States. It's also heard when Lieutenant Colonel Moore discovers Jack's dead body on the battlefield. From this point on, Jack's theme dominates the score. A parallel set of scenes crosscutting between Vietnam and the United States follows the discovery of the dead Jack, with the theme in a strings-only, lushly contrapuntal version scoring Moore sitting quietly with Jack's body in Vietnam and Moore's wife, Julie, delivering the news to Jack's wife, Barbara. The score falls briefly silent when Julie tells Barbara of Jack's death, creating an ABA (score/dialogue/score) structure that allows for Jack's theme to reenter as a music now expressly tuned to the grieving widow's reaction. Having used Jack's theme in targeted fashion during the film's narrative, Glennie-Smith uses it pervasively for the film's long, slow, final movement. The last battle sequence—a highly stylized, tremendously violent montage—is a slaughter of the Vietnamese army by Moore's company, enhanced by crushing American air power. In the immediate aftermath, a tight mesh of intense, slow-moving, consonant strings—again echoing Barber's *Adagio for Strings*—accompanies the sight of an enormous pile of bloodied Vietnamese bodies left on the field of battle, an image immediately juxtaposed with Jack's beautiful corpse being carefully returned to the States. Jack's theme begins and continues, with a few interruptions, for the next ten minutes. Elegiac string music, consonant in harmony and contrapuntal in texture, pours out of the soundtrack as Moore finds himself speechless before the just-arrived reporters and during a final conversation between Moore and reporter Joe Galloway about how the difficult story of the battle must be told. The battlefield dissolves to Moore's house in the States. After a brief pause, cheaply dramatizing Moore's wife's apparent fear her husband may have been killed, Jack's theme restarts and plays on and on in an energized major-mode version as Moore is reunited with his family. Jack's wife, a Vietnamese soldier's widow, and Galloway (crying as he types) appear in turn—each reflecting on their losses—with Moore and Galloway heard in voice-over: all get Jack's theme for background. And Jack's theme, this time for solo trumpet, continues as Moore visits the Vietnam Veterans Memorial, as described in chapter 1. When the end titles begin rolling, "Jack's Theme" acquires a text, written by director Randall Wallace and sung by the US Military Academy Cadet Glee Club. This hymn, "Mansions of the Lord," served as the recessional for the funeral of President Ronald Reagan two years after *We Were Soldiers* was released and can still be heard at Veterans Day events across the nation. Glennie-Smith's theme, set for strings, anoints the young soldier Jack and scores the grief of his loved ones. Restrained brass, and then reserved but strong voices, transfer this music for one real-life soldier into a generalized memorial for all the fallen—and outside its original context into a hymn for general use in the post-Vietnam era.

The original scores for Eastwood's Iwo Jima diptych form a strong element of expressive overlap between the two films. Eastwood is credited as composer on *Flags of Our Fathers*; Eastwood's son Kyle and Michael Stevens are credited together on *Letters from Iwo Jima*. Both scores are modestly scaled: generally at a low dynamic, reserved in tone, lyrical, and melancholy. Eastwood father and son are featured in separate interviews about the music on the *Letters from Iwo Jima* score CD. The former speaks of not wanting music to "run the show," favoring instead a score where "less is best." But Eastwood also explains how he likes to "cut to the theme," often beginning work on a score during the editing process. By this method, music track and image track grow in a gradual and reciprocal fashion. Eastwood notes how they did not want the "cliché" of a Japanese theme reminiscent either of Hollywood films like *Sayonara* (1957) or of Kabuki theater. Kyle Eastwood adds that they avoided "bamboo flutes," a choice that ruled out a kind of Asian "exotic" veil music heard in many Vietnam PCFs (see chapter 8). Instead, the *Letters from Iwo Jima* score uses the essential solo sounds of the *Flags of Our Fathers* score: trumpet and piano. Eastwood claims the theme for *Letters from Iwo Jima* grew slowly, taking "some clue" from an authentic patriotic song—a Western-style tune with vaguely "Asian" melodic turns, by musical analogy expressing Japan's Western-style military might and deeply traditional military values—sung on a radio broadcast to the island by children in Japan, a haunting moment of diegetic performance that echoes via loudspeakers through the caves where the Japanese soldiers are preparing to fight and die. In the interview, Eastwood says of this process, "Pretty soon there was a little score." This is an understatement. Music of a similar kind is present in both films as a central expressive element granting a similar if differently shaded pathos to American and Japanese soldiers alike. The choice of instruments and tune in *Letters from Iwo Jima* situates the film's Japanese characters in an unusual musical space, free of both Hollywood stereotypes and period touches. (Chapter 9 discusses the *Flags of Our Fathers* score.)

Almost Continuous Scores *(Apocalypse Now, The Thin Red Line, Black Hawk Down, The Kingdom, Green Zone, Act of Valor)*

Distinguishing music-laden films from films with almost continuous scores involves assessment of sheer quantity—second-by-second accounting for the musical thread of the soundtrack—and of the default nature of the mix. Films with almost continuous scores use music pervasively either to maintain a kind of dream state—as in *The Thin Red Line*—or to support a continuous action narrative—as in *Black Hawk Down, The Kingdom, Green Zone*, and *Act of Valor*. As is often the case, *Apocalypse Now* proves unique among PCFs in its use of almost continuous music blending into an equally continuous atmospheric effects track.

The score for *The Thin Red Line* almost never stops. In the film's 170-minute run time, musical silence is observed only sixteen times. Pauses in the score last on average about two and a half minutes; half are shorter than two minutes. The score as a whole encompasses more than two hours of music. (The score album lasts but 82 minutes. The differences between film score and score album are many.) The default mode of the mix includes music, typically quite prominent, often in a context of near-diegetic silence and/or voice-over. Director Terrence Malick's musical practice across his output informs this larger aesthetic choice, but quantifying the music in *The Thin Red Line* helps account for the film's overwhelmingly lyrical effect—which enhances the philosophical and "universal" conceit of the movie—and draws attention to the film's essential need for music as a formal support for its very structure, an aspect of this much-studied film that has gone undiscussed. (Chapters 5 and 9 consider how music formally and expressively supports voice-over in this film.)

A group of twenty-first-century PCFs fall within a larger trend toward continuous scores in action films. In these films, beat-driven music, often drawing on electronically generated or synthesized sounds, plays almost continually, as in a video game. A driving beat stops and starts, accelerates or drops back, in relation to narrative events or the (often short bursts of) dialogue. Music paces the action in these relentlessly moving, tightly structured, often complex narratives, which are marked visually by jerky and chaotic camerawork and disjointed and rapid editing. *Black Hawk Down, The Kingdom, Green Zone,* and *Act of Valor* are all marked by a virtually continuous beat. One distinguishing feature of these films' modern warfare scores relative to more generic action films is that they all make reference to the key PCF musical trope: elegiac music on the model of *Platoon's* use of Barber's *Adagio for Strings.* They do so to greater or lesser extents, but all make a feint toward this more serious, reflective musical register. (Chapter 9 takes up these beat-oriented scores as part of a larger discussion of meter in PCF music.)

The soundtrack for *Apocalypse Now* consistently blurs the distinction between scoring and sound effects, raising a set of unique issues around the film. There are, of course, many instances of music alone filling the soundtrack. The opening sequence using the Doors' "The End" establishes this approach to the mix as a defining strategy of the film's representation of Vietnam as a fever dream, a sort of madness that requires what Coppola and screenwriter John Milius called "hard rock" music. (Chapter 4 considers how rock music of a specific type serves this theme in the film, especially in script drafts including unmade battle sequences.) But more often, the mix layers music and sound effects, fading from one into the other seamlessly. For example, after Lieutenant Colonel Kilgore's wistful line "Someday this war's gonna end," two distinct, unrelated musical elements—"jungle" drums sounding a stereotypically primitive beat and a tangle of glass harmonica–like quasi-melodic lines—are layered with a voice-over by Willard and the casually overheard voices of the other men in the boat. The men's voices gradually assume prominence—we begin to register Chef talking about mangoes—as the musical elements cross-fade into an extremely active jungle-sounds mix of birds and insects, with distant thunder or bombing, on the effects track. These noisy jungle sounds, sustained in terms of texture and volume, continue as Chef and Willard talk while moving through the jungle looking for mangoes, only falling silent when Willard senses something, perhaps the enemy, nearby. The jungle momentarily falls quiet and a very high-pitched, sustained sound comes in as Willard moves forward on alert. This high sound could be an insect: it's not far from the unidentified jungle sounds heard just before. But as other sounds come in—sounds that hint at musical origins from one of the four synthesizer artists engaged for the score—the mix tilts toward the musical.

The soundtrack falls almost completely silent just before a tiger leaps toward the pair, its fantastic roar more frightening than its visual presence. Chef and Willard flee toward the sound of gunfire from Clean on the .50 cal machine gun and the boat's loud engine, an alternate set of atmospheric, slightly musical sounds—especially the rhythmic bursts of automatic weapons fire—which displace the continuously activated sound field of the jungle, continuing the soundtrack's strategy of constant noise of some sort almost all the time. As Chef calms down, the score noticeably sneaks in on a return to Willard: a synth hum displaces the engine's hum and a meandering flute cues another ruminative voice-over, beginning with the words, taken from Chef's frantic cries, "Never get out of the boat."

Apocalypse Now's layered, dense, but transparent mix—consistently blurring the line between distinctly musical sounds and effects textured in a musical manner—was likely encouraged by the then-still-new technical and expressive challenge of crafting a surround-sound mix, which in the case of *Apocalypse Now* was conceived for a so-called Quintaphonic speaker setup, drawing on the aesthetics of rock concerts and records.[26] Keeping all speakers engaged in the mix seems to have, in part, influenced the film's almost continuously activated score and effects tracks. The soundtrack, in an almost constant state of hum, burble, rattle, and shush, is virtually never clear of noise, and much of the original score, narrowly understood as recognizably musical sounds created specifically for the film (perhaps dated to contemporary ears due to the dominance of synthesizers) similarly traffics in sound fields, atmospheres, and textures, often without apparent meter or melody. As David Rubinson, the music producer on the film, commented in 1979: "It's not picture . . . dialogue . . . effects . . . music like it usually is in film. It is all integrated so the effects become the music, the music becomes the effects, explosions become the music. A synthesized helicopter becomes the bass line to a piece of music. The bass line evolves until it becomes a motor boat pulling out of a harbor. The helicopter rotors whir until they become a cello line. It is a *sound* environment."[27]

In such a context, scenes without music or atmospheric sound effects take on a heightened quality. Perhaps the central episode of the film, the most direct in its portrayal of combat in Vietnam, is the encounter between the patrol boat (PBR) and a sampan: when the Americans massacre a group of Vietnamese civilians for no apparent reason, largely, it seems, out of the younger American soldiers' fear. This scene is lacking in the 1975 script and must have been added during the shoot, when Coppola was rewriting the film nightly. It adds nothing substantial to the plot but does take *Apocalypse Now* out of its psychedelic default mode. The scene proper begins when the

PBR is secured to the sampan and Chief cuts off the boat's engine. The mix falls disarmingly quiet at this moment. For the duration of the violent and emotional scene, which climaxes with Willard's point-blank shooting of an injured Vietnamese woman, *Apocalypse Now*'s otherwise almost always activated score and effects track falls silent, withdrawing from the viewer the film's default sonically surreal means to lean back from the narrative. Indeed, the end of this scene is not musical. Willard, having shot the woman, sits on the edge of the boat and the image fades slowly to black. Silence reigns for several seconds in the darkness before the film continues. This is a rare moment in *Apocalypse Now*, articulating large-scale form with a fade-out—rather than a cross-fade into the next episode—and offering time for reflection on the narrative that is undirected by music or effects. (On a more practical note, the fade-out may be the remainder of a planned intermission. In that case, Coppola intended to send his audience into the break without music, a stark choice given the film's richness of musical moments to this point.)

In a 1996 interview about his work on *Apocalypse Now*, Walter Murch advanced an extended metaphor about sound and meaning in film:

> Think of a spectrum of sounds as you would think of a spectrum in a rainbow. At one end of the rainbow you have red, at the other end you have blue. There is, at least conceptually, a kind of equivalent of that in sound where you have at one end of the spectrum sound that has a high degree of meaning in it. The most obvious example of this is dialogue. When you have dialogue in a film it is being conveyed by sound, but what is being conveyed is a code, and in order to understand what film-makers are talking about you have to understand that language. So the shell which is put in your hands is sound, and you, through your ability to understand language, crack open that shell and consume what is inside which is the meaning of the words. Let's say that that is the blue end, the ultraviolet end. At the far end, equally opposed to that is infrared—music—which has very little code, very little language, very little specific meaning, but tons of emotion, tons of manner, and an unrivalled ability to communicate mood to the audience. That gives us two ends to the rainbow, but obviously it's never quite as clear cut as that. There are elements of code that work their way into music, and by the same token, how somebody says something is very strongly influenced by the audience in that they decide about the meaning of what is said, so there is an element of music in any form of speech. But in between is this rather vague area of just pure sound: it's not dialogue any more, but it's not really music. It's what would be conventionally thought of in terms of sound effects, but it can float rather

independently from one end to the other. Sometimes sound can become almost pure music; it doesn't declare itself as music which is actually one of its advantages because it can have a musical effect on you without you realizing it, but at the other end it can sometimes deliver very discrete packets of meaning.[28]

Parts 2, 3, and 4 of this book respectively unpack the coded language of dialogue, the middle ground of effects, and the emotional and mood impact of music. As Murch emphasizes, all three sorts of sound can float rather freely: the cadence of dialogue and the rhythm and timbre of effects can work like music; the sounds of weapons near and far can spell out plot developments; music can speak quite specifically within a context of generic registers. The PCF soundtrack offers a soundscape of war that utilizes all three sonic elements toward a meaningful whole.

J. Martin Daughtry offers a reciprocal if less aesthetic view of all possible sounds in his 2015 book *Listening to War: Sound, Music, Trauma, and Survival in Wartime Iraq*. Daughtry coins the term *belliphonic* to refer to "the spectrum of sounds produced by armed combat."[29] He breaks the belliphonic into categories roughly parallel to the three elements of the soundtrack: the sounds of weaponry, motorized vehicles, and generators running in the absence of the power grid; the way soldiers talk about sound; and music, understood as part of the experience of war and not an autonomous sonic realm, even though Daughtry's soldiers and civilian informants "desperately [want] it to be."[30] Daughtry defines various strategic listening practices that soldiers and civilians in Iraq engaged in to survive physically, mentally, and emotionally. He details the "informationally rich signals" of the battle zone, parses "the acoustic signatures of missiles, mortars, and gunfire," and describes his "belliphonic auditors" as "people who experience war through their ears; never exclusively, but importantly, and often, as I will argue, centrally."[31]

Allowing for the fundamental, indeed existential, difference between soldiers and civilians living in a war zone and an audience watching a war movie in the theater or at home, I would suggest that Daughtry's notion of actual belliphonic sound is useful for this study of simulated belliphonic sound. PCF audiences experience simulated war "through their ears" and learn to read "informationally rich signals" while following film plots and, potentially more richly, partake vicariously of the soldier and veteran experience. They do so, of course, in a context where director and mixer exercise sovereign control over sound and where the statutory regulation of amplified sound limits the volume to physically safe levels. Daughtry explores at length the power of loud sounds to cause trauma to body and mind.

Explosions in a film—no matter how loud—only ever hint at the physical power of war's explosive sounds. Furthermore, the control over sound provided by the technologies of film allows the director and mixer to create soundscapes of total or diegetic silence, moments of respite when all sound falls away or music alone floods the heard world. At such moments—and there are many in the PCF—filmmakers anticipate that their audience "desperately wants [music] to be" "an autonomous world unto itself."[32] Film can grant that wish to movie audiences. The real-world belliphonic offers no such refuge.

PART II

Dialogue

3.   Soldiers' Talk

Fighting men talking—"while it did not hit any stratosphere of wit"—fills the pages of reporter Richard Tregaskis's *Guadalcanal Diary*, published in January 1943 and rapidly turned into a genre-defining combat movie released in October of the same year.[1] At the film's opening, Tregaskis, in voice-over, introduces the main characters sunning themselves on deck, saying, "The favorite occupation, as usual, is shooting the breeze, exchanging scuttlebutt." Eavesdropping on soldier talk—on men speaking freely among themselves—provides one attraction of the combat film, a decidedly homosocial genre. But of course the conversation among the Marines in *Guadalcanal Diary* stays safely within the limits of the Production Code. Even in his book, Tregaskis couldn't print the word *fuck*. He did, however, note its popularity in an offhand moment: "The Sheik only chuckled. 'F——— you, Mac,' he said, indulging in the marines' favorite word."[2]

In his 2004 book *Generation Kill*, reporter Evan Wright similarly captures the voices of young Marines in the 2003 invasion of Iraq talking endlessly about anything and everything. Their words, however, are completely uncensored. Wright's book presents an array of distinctive characters composed in large part by their speech. The HBO series *Generation Kill* (2008), protected by its cable context (and excellently cast with mostly unknown actors), allowed these guys to say anything and everything as well. Their endless chatter is the primary pleasure of the series, devoted as it is to combat frustration.[3]

This chapter selectively surveys dialogue in the PCF from several angles: content (what is said), tone (how soldiers and the actors playing them speak), and repetition (how words and phrases, sometimes drawn from military culture, act as refrains shaping select films' and cycles' verbal structure).

In *Action Speaks Louder: Violence, Spectacle, and the American Action Movie*, Eric Lichtenfeld includes but one reference to dialogue in his list of the basic ingredients of post-1970 Hollywood action films: "one-liners," he notes, prove a crucial verbal identifier for the genre.[4] Consider the flatly delivered "I'll be back" spoken by Arnold Schwarzenegger in *The Terminator* (1984). This three-word sentence captures the surprisingly satisfying lack of dimension to the Terminator as a character and marks the acceptance—indeed, pleasures—of limited expressive range in action film actors' performances. The phrase itself promises immediate action and a sequel—he will indeed be (right) back and he'll say it again (in subsequent films). And repeating the phrase out of context, impersonating character and actor, efficiently invokes the entire world of the franchise. Not every action film or franchise arrives at the verbal concision of "I'll be back," but most aspire to the phrase's stripped-down aesthetic and branding potential. At the very least, action filmmakers look for the snappy line preceding the kill or the terse post-kill riposte—as in "Consider that a divorce," spoken by Schwarzenegger after shooting his fake wife in *Total Recall* (1990). The practice is so tightly wed to the genre that the pop culture website Gawker posted a nine-minute video of the one hundred greatest action-movie one-liners. All but ten films included date to the era of the PCF, and these films form the background context of the subgenre.[5] But only one PCF made Gawker's compilation: *Apocalypse Now* (1979) is represented by Lieutenant Colonel Kilgore's line "I love the smell of napalm in the morning." In the film, Kilgore follows this sentence with a story of wandering on a hillside destroyed by a napalm strike. He didn't see a single enemy body—all were burned to ash—and took pleasure in the pervasive odor of gasoline (a primary ingredient in the weapon). Kilgore ends his story with the valediction—"It smelled like victory"—and a rueful exit line—"Someday this war's gonna end." In context, and as delivered by Robert Duvall, Kilgore's lines work nothing like generic action-film taglines. PCF dialogue typically avoids the memorable one-liner. The men in PCFs seldom talk like action-movie cartoon characters. And when they do, the colorful language of action heroes is put to meaningful work toward the subgenre's larger goals.

Consider for example the word *motherfucker*. Several action-film one-liners incorporate this liminal if common profanity, sometimes for quasi-comic effect: "Die screaming, motherfucker!" in *The Long Kiss Goodnight* (1996); "I have had it with these motherfucking snakes on this motherfucking plane!" in *Snakes on a Plane* (2006); and most famously, "Yippee ki-yay motherfucker" in *Die Hard* (1988) and its sequels. Edging nearer to the PCF, director William Friedkin's 2000 military drama *Rules of Engagement* suggests how the word *motherfucker* operates in more serious screen contexts. The film's story turns on a demonstration against a US embassy in the Middle East. The commander of the Marines protecting the embassy, played by Samuel L. Jackson—a star with a special relationship to virtuoso profanity—orders his men to fire on the crowd of demonstrators with the cry, "Waste the motherfuckers!" The crowd is massacred, including children, causing an international incident. Jackson's court martial trial serves as the climax of the narrative. Jackson is exonerated, but not before a moment of bemused despair between Jackson and his Vietnam buddy and defense lawyer, played by Tommy Lee Jones, over courtroom reaction to Jackson's use of *motherfucker* at the height of the fight. The two combat-tested Marines hear it as a battle cry while the military top brass, public, and press hear it as a sign of extreme aggression, unthinking force, and prejudicial intent. The latter interpretation is normative in the PCF.

Indeed, the word *motherfucker* runs a curious path through the PCF, connected consistently to extreme characters or moments. In *Casualties of War* (1989), Meserve—the sergeant behind the kidnap, rape, and murder of a young Vietnamese woman, played by Sean Penn—yells "Get some, motherfuckers, get some" while firing a SAW (a lightweight machine gun) at the enemy. He smiles in grotesque pleasure, a look of violent ecstasy repeated by Penn whenever his character fires on automatic. In *Jarhead* (2005), while watching *Apocalypse Now*'s helicopter attack scene with a theater full of shouting Marines—the film's version of the pre-deployment war movie festival described in the book—Swofford mutters to himself, then shouts out, "Shoot that motherfucker," all while touching his face and head in a quasi-sexual frenzy. Perhaps most menacingly, in *Platoon* (1986), after a booby trap kills several men in grotesque fashion and the platoon discovers Manny, a missing soldier, garroted by the enemy, Tom Berenger as Sergeant Barnes—in the words of Stone's script—"says it for everyone, 'The motherfuckers.'"[6] Berenger's menacing muttering of the word verbally kicks off the central village atrocity scene in *Platoon*.

Shouting obscenities while firing is not, of course, unknown among real soldiers in combat. The journalist Bing West documents several such moments in *No True Glory: A Frontline Account of the Battle for Fallujah* (2005), which was for a time under development for making into a movie.[7] In the midst of battle, one American cried out—as if quoting a movie—"Yeah! Get some! Come on, motherfucker! Get some!," then "burst out laughing."[8] (Evan Wright defines "Get some!" as the "unofficial Marine Corps cheer . . . [expressing the] excitement, the fear, the feelings of power and the erotic-tinged thrill that comes from confronting the extreme physical and emotional challenges posed by death, which is, of course, what war is all about;"[9] one might add the concurrent thrill of killing, which is also what war—and war movies—are about.) But the limited, always thematically troubling, representation of soldiers enjoying the pleasures of battle while shouting "motherfucker" in the PCF signals a persistent angle of the subgenre: if soldiers are to be presented as sympathetic figures to an audience other than young men, then the way they speak must be nuanced. Moderating soldiers' language proves one way to do this.

In contrast to action film norms, the overall tone of the language in PCFs is restrained and plain. These are not especially articulate or clever-speaking fellows. They generally don't shout obscenities while firing on automatic or sling one-liners along with their weapons. With some notable exceptions, the PCF is defined on the dialogue track by a reserved sort of male speech and verbally contained central characters. Critics respond to this lack of verbal bravado and irony: "Amazingly, when [Ice Cube in *Three Kings* (1999)] blows up the aforementioned helicopter, he does not follow the explosion with a clever quip! Perhaps we're finally getting somewhere."[10] One way to observe PCF dialogue norms being set is to compare a film's source materials and draft and published scripts with the finished screen version. Here, evidence for the precise calibration of a film's verbal diction can be discovered.

Differences between the scripts and the finished film in the case of *Saving Private Ryan* (1998) show how the central character of Captain Miller was altered at the level of action and speech, making him softer, more sympathetic, and less generically heroic, matching the part closely to actor Tom Hanks's persona. Robert Rodat's original script introduces Miller in the Higgins boat heading toward Omaha Beach this way: "Relaxed, battle-hardened, powerful, ignoring the hell around them. He smiles, puts a cigar in his mouth, strikes a match on the front of DeLancey's helmet [*sic*] and lights the cigar."[11] In the film, Spielberg emphasizes Miller's uncontrollably shaking hand as he lifts a canteen to his lips. During D-Day, Rodat's Miller repeatedly exposes himself to enemy fire, seemingly fearless of the slaughter around him. Rodat wanted to show Miller and his men scaling the cliffs at Pointe du Hoc, a heroic Army Ranger mission which, while the casualties were high, is still celebrated as a memorable act of battlefield skill. Spielberg chose instead to represent the landings on Omaha Beach, the bloodiest, least overtly heroic sector, where the historical narrative has centered on simple survival against seemingly impossible odds.

Interestingly, Rodat's Miller, who has already been awarded the Congressional Medal of Honor, supposedly enabling him to sass any officer as he wishes, takes a strongly negative view of the mission to find Ryan, calling it, to his commanding officer no less, "bleeding-heart crapola from three thousand miles away." All such talk is cut in the film. A further cut exchange that would perhaps have offended the targeted audience of older veterans (watching with their wives) with its use of the word *fuck* and sexual innuendo suggests how *Saving Private Ryan* might have been an earthier, less polite film:

REIBEN: Yes, sir, as a final note, I'd like to say, fuck our orders, fuck Ramele, fuck the cheese capital of France and while we're at it, fuck Private James Ryan.

MILLER: I'll make a note of your suggestions but I'll leave that last one to you, especially if he's already dead.

The men wince and laugh.

Hanks's Miller is reserved in his opinions about the mission, never quite undercutting the fairly absurd premise of the plot. Hanks's subtlety—or blandness—allows the film to come off as perhaps more reverent than it is.

Paul Greengrass's dialogue sets a properly serious tone in *United 93* (2006). The published script includes the following lines—none of them heard in the film—that signal generic action-film humor, irony, and bravado:

- After the first plane hits the World Trade Center, Ben Sliney, new head of the air traffic control center, brings the attention to himself, complaining "On my first day."

- After a stewardess is brutally killed on United 93, a message from United's dispatch center asks, "Can dispatch be of any assistance?"

- Frustrated at the lack of success scrambling planes to defend the Eastern seaboard, a controller at NEADS says, "If I could shit an F16 right now I would."[12]

Inclusion of any of these lines would have activated a kind of movie dialogue awareness and movie character bravado that Greengrass ultimately determined was inappropriate for *United 93*.

Tone of voice is crucial to the dialogue aesthetic of the PCF. Remaining with *United 93*, in the script Greengrass assigns the line "Let's roll" to "A Voice." In the film, the line is spoken by Todd Beamer, the passenger to whom the line is attributed (Beamer's wife, Lisa, wrote a best-selling book about him titled *Let's Roll!: Ordinary People, Extraordinary Courage*).

Beamer in *United 93* says "Let's roll" in a loud whisper, with a sense of urgency but no heroics. His words do not initiate the passengers' counterattack—as they might in an action flick—and go generally unheeded. The moment passes quickly in the film's dry presentation of events on the plane. President George W. Bush famously used "Let's roll" as a rallying cry in his speech of November 8, 2001: "We will always remember the words of that brave man, expressing the spirit of that great country. . . . We have our marching orders. My fellow Americans, 'Let's roll.'"[13] *United 93* offers a riposte to any use of Beamer's words as "marching orders." The tone and delivery of the line, deemphasizing the day's most famous words, preserve the film's larger documentary tone, simultaneously keeping it from slipping into jingoistic territory and allowing viewers resistant to Bush's use of the phrase to embrace the film: registering the sacrifice of the passengers without embracing them as the first soldiers in the GWOT. Moderated tone plays a key role in keeping the door open to such readings.

Minimizing aggressive talk is one tack toward finding an acceptably sensitive yet still masculine tone of voice. The script for *Flags of Our Fathers* (2006) unsurprisingly omits Sergeant Mike Strank's words in the book, shouted while leading his men into battle: "Let's show these bastards what a real banzai is like! Easy Company, charge!"[14] It's difficult to imagine actor Barry Pepper's laconic and caring Strank ever saying such things. The PCF is invested in more modest speech, in "heroes who do not pose as such."[15] But there are limits on the sensitive side as well. For example, only two PCFs construct a leading character around religious faith. Hal Moore in *We Were Soldiers* (2002) is shown praying several times, once with his many small children. For good measure, Moore undercuts an initially pious prayer with his subordinate Jack by closing with a hyper-masculine, perhaps tongue-in-cheek request for God to help them "kill the heathen sons a bitches." Jack gives Moore a quizzical look—offering a point of distanced identification for audience members who want it. Another quote from a flag raiser in the book *Flags of Our Fathers* (2000) not heard in the film is James Bradley's nightly prayer, "Blessed Mother, please help us so everything turns out all right."[16] It would have pushed Ryan Phillippe's portrayal of Bradley out of an ambiguously gentle range to make him a devout Catholic seeking help from the Virgin Mary. Indeed, the only other praying man in the PCF is Captain Staros, played by the soft-spoken Elias Koteas, who vaguely asks for guidance the night before battle in *The Thin Red Line* (1998). Staros's care for his men leads him to defy orders to send them into withering fire and certain death. He is relieved of command and exits the film intoning that his men are "like my sons," then that they "are my sons." Staros takes the gentle soldier of the PCF to an extreme. He never says he's their father; his actions suggest he could be their mother.

Contrasting tones of voice and verbal personas also work to privilege certain kinds of leadership within the PCF and to frame dialogue between unequal ranks as a meaningful space for the expression of a film's core issues. The contrast between excited, usually inexperienced, soldiers who shout and experienced soldiers who speak in measured tones recurs in several films, at times as a matter of disciplining young soldiers to remain calm under pressure. While calling in an air strike, the young radio operator in *Hamburger Hill* (1987) shouts, "Blow the shit out of them," to which the voice on the radio replies, "Use proper radio procedure." (Advising Stanley Kubrick on the characteristic sounds of Vietnam, Gustav Hasford—war journalist and author of the 1979 book *The Short-Timers*, the novel on which *Full Metal Jacket* [1987] is based—noted the monotone voices on the radio.)[17] *Black Hawk Down* (2001) develops this contrast on several levels. At the start of the film, the young Ranger Matt Eversmann shouts excitedly into his headset about Somali militias firing on civilians; the more experienced Black Hawk pilot surveys the same scene and, as described in the script, "calmly speaks into his headset" and says "I don't think we can touch this."[18] Later, the go signal for the mission is spoken clearly and distinctly by a mid-level commander: "Irene, I say again Irene." A Black Hawk pilot, starting the rotor to turning, gives out with a gleeful shout, "Fuckin' Irene!" General Garrison—introduced in the script as "laconic. Steady." and played with characteristic reserve by Sam Shepard—remains completely calm through all the day's events, which he observes from the tactical operations center (TOC). Several abrupt shifts from the heat of the battle on the streets to the quiet of the TOC come as shocks: Garrison's abiding professional demeanor can appear detached. Only in his final scene does Garrison betray emotion, and then by way of actions rather than words. Passing through the base hospital, he tries to wipe up a pool of blood on the floor with some paper towels, only succeeding in making the stain larger. The script mentions Garrison using a mop, a less dignified tool than he uses in the film.[19] *Black Hawk Down* carefully calibrates Garrison's masculine effectiveness as the leader of a failed mission.

Black Hawk Down features one character who, in the original scripts, acted and spoke like an action-flick hero. Early scripts had the Delta Force tough guy "Hoot," played by Eric Bana, deliver several one-liners:

- "Don't even think about it" said to a Somali reachii ´ ˉ weapon.
- "That's what God made bullets for" to Garrison.

- A planned sequence that might have garnered big action-movie laughs had Hoot attack a desktop PC, saying "Control"—then rack his shotgun—"Alt"—then blast a hole in the computer—"Delete."[20]

While Hoot acts with extreme effectiveness in combat—anticipating the SEALs of later films—his action-movie pose gets reduced to hard looks, sunglasses, and a humorous moment early on when he wiggles his trigger finger and says that's his "safety." Instead of offering taglines, Hoot offers advice in three pointed conversations with Eversmann, who is commanding men in battle for the first time. Distributed across the film—one before the fight, one at a low point, one at the close—Hoot acts as counselor to the young man, and his words could be a description of the PCF as *Black Hawk Down* would shape it:

- "Know what I think? Don't really matter what I think. Once that first bullet goes past your head—politics, and all that shit—-just goes right out the window."
- "So you're thinking. Don't. [beat] Cause Sergeant, you can't control who gets hit, or who doesn't. Who falls out of a chopper, or why. It ain't up to you. It's just war."
- "They won't understand, it's about the man next to you . . . and that's it . . . that's all it is."[21]

Hoot, an understated voice of the subgenre for the twenty-first century, offers a modest reminder of what's at stake: a masculinity based on cold knowledge of combat (and little else) that doesn't boast of but instead fiercely speaks of the bonds between men. (*The Thin Red Line* similarly distributes three talks between Welsh and Witt across the film as points of reference. These talks touch on similar questions but in more abstract, philosophical terms.) Only one tagline for Hoot survived. Just before the final action sequence, as the sound of gunfire grows, he looks at Eversmann and says, "Well, shall we?" Hoot could have completed the thought and quoted Kilgore in *Apocalypse Now*, whose words "shall we dance" kick off the helicopter attack to Richard Wagner's "Ride of the Valkyries." In both cases, battle as a dance for men and machines is initiated by a spoken invitation that shifts the mix toward a predominance of music and effects.

Supremely competent, indeed evidently fearless, professional soldiers become the norm in post–*Black Hawk Down* PCFs. Toning down their spoken rhetoric remained a consistent strategy. For example, Staff Sergeant Will James, the bomb tech at the center of *The Hurt Locker* (2008), was to launch a mission by saying, "Let's rock 'n' roll." In the film, James utters a more moderate "Let's do it." His other one-liners are thrown away. "If I'm gonna die, I wanna die comfortable" as he tosses away his headphones, and a post–bomb defusing cigarette lit with the words "That was good," don't come off as movie posturing, although in a different context they might read as classic one-liners. Jeremy Renner's Oscar-nominated performance is central here. Indeed, the PCF has benefited from both attracting highly skilled, chiefly dramatic actors—of Hanks's caliber, for example—to the action genre and the persistent use of unknown actors who do not bring existing personas or, in the case of *Act of Valor* (2012), any acting experience to the screen. (For example, consider how the tough-guy action persona of Vin Diesel inflects his gentle performance in *Saving Private Ryan* only in retrospect, now that we've experienced his subsequent career.)

James's (and Renner's) moderate demeanor is enhanced by the appearance of gung-ho minor characters, for example a blowhard captain who calls James "hot shit" and "a wild man." The ridiculous, cartoonish aggression of the captain makes the sober trio at the center of *The Hurt Locker* seem more mature by contrast. (Surrounding a nuanced, understated masculinity with broad caricatures is an old Hollywood trick: for example in his early films Fred Astaire was made more manly by supporting casts of effeminate male stereotypes.) Several PCFs develop the trope of the overexcited superior officer whose words give away his unreliable character: Sobel's cries of "Hi-ho Silver" in *Band of Brothers* (2001) and Captain America "spazzing out on our comms," as a junior officer says of an out-of-control superior in *Generation Kill*. The angry rants of Colonel Tall, played by Nick Nolte, in *The Thin Red Line* are contextualized by the character's introduction off the line of battle, when voice-overs offering access to Tall's inner life situate his shouting as rooted in insecurity.

There is, of course, one great exception to the PCF practice of moderating generic action-film speech: the drill instructor or sergeant. Drill instructors appear in *The Boys in Company C* (1978), *Full Metal Jacket*, *Jarhead*, *The Pacific* (2010), and *American Sniper* (2014). The cinematic origins of a shouting drill instructor reach back to the 1957 film *The D.I.* Actor Jack Webb—who directed and produced it and appears with a cast of active-duty soldiers deployed to the film as actors (similar to *Act of Valor*)—made ample room in his performance for the drill sergeant's offstage humanity. Military discipline is presented as a learned performance: to be a soldier, first learn to act like a soldier—but never to the loss of depth as a person.

Lee Ermey's drill instructor in *The Boys in Company C* also makes plenty of room for humanity. We see him offstage showing genuine concern for the lives of his men. The innovation in this film is the excessive profanity, allowed by the still-recent lifting of the Code and remarked on by almost every reviewer: "It seems like every-other-word is a swear word. While this may be Vintage Marine Talk, the cumulative effect is soon wearying and takes an unnecessary toll on an audience."[22] It depends, of course, on the audience. *The Boys in Company C* seems made for a male crowd: *Variety* noted that it "[plays] at the enlisted man's level."[23] It introduces the drill instructor as the master of virtuoso crude talk. But it does so, crucially, while preserving a level of casual, modest talk among men, including the D.I. himself. *Box Office* called Ermey "a former Marine drill instructor who acts with a naturalness to be admired."[24]

It's hard to imagine a reviewer making the same comment about Ermey's more famous performance as Drill Instructor Hartman in *Full Metal Jacket*, where he gives a completely stylized performance with no offstage moments. The soldiers he trains are similarly stylized. There are no real human characters in Kubrick's film. (Kate McQuiston argues with regard to spoken language in the film that Kubrick "walks a frightening line between caricature and realism.")[25] And as *Newsweek* noted, "There's hardly a straight English sentence in the film; from the Rabelaisian vituperations of the drill instructor to the street-hip, rock-and-roll happing of the grunts, it's as if language itself has fragmented along with the values that no longer hold it together."[26] (Hasford encouraged this stylized approach to dialogue, arguing in preproduction notes to Kubrick that Marines share a "specific" and "particular language" identifying them as Marines.)[27] Ermey himself thought his performance crossed a line of verisimilitude: "Any drill instructor is an actor to some extent. Nobody's that nasty."[28]

Tough talk in the PCF is typically held in reserve as a weapon for rare, sudden, expressively sharp use. A choice moment from *Zero Dark Thirty* (2012) reveals how a single line of boastful, in context jarring, profanity can efficiently suggest that an aggressive spirit animates the efforts of the sober professional American soldier. This abrupt eruption of strong language, delivered by a serious actor in a tightly controlled manner, situates intelligence analysis as both the GWOT equivalent of combat and a domain of contemporary warfare where women might be the equals of men. The scene is a high-level meeting where CIA Director Leon Panetta, played by James Gandolfini, is being briefed on the compound in Pakistan that analysts believe to be Osama bin Laden's hiding place. CIA analyst Maya, played by Jessica Chastain, injects herself into the conversation—where she has no standing to speak—by correcting her superior's answer to one of Panetta's questions. Panetta turns to Maya and asks, "Who are you?" Posed beside a framed American flag, she replies flatly, "I'm the motherfucker that found this place, sir."

War films have always taught audiences—chiefly young men—how war works. Learning the lingo and litanies of soldiering is part of the experience gained from watching these films. A basic plot trajectory provides ample opportunity for such education. Cherries, FNGs (fucking new guys), or replacements—slang terms for inexperienced soldiers—show up prominently in *Platoon, Hamburger Hill, Casualties of War, Saving Private Ryan, Band of Brothers,* and *The Pacific.* The cherry needs things explained and defined—for example, Private Upham needs to learn the meaning of "fubar" in *Saving Private Ryan*—and their learning curve becomes that of the audience. As veteran Jim Carabatsos notes late in the script for *Hamburger Hill,* "The new guys are picking up the cadence and slang of the 'Nam."[29] Films with extended training sections, such as *The Boys in Company C, Full Metal Jacket, We Were Soldiers,* and *Jarhead,* provide a group immersion experience: the audience learns with the men.

The "new guy" plot is less accessible in films about the all-volunteer army and is totally out of place in GWOT films about elite warriors: professionals know their stuff. But professionals have rituals of membership. The hinge point of *Lone Survivor's* (2013) two-part plot—the transition from preparation to execution of the doomed mission—is finessed by a "new guy ceremony." A young SEAL played by Alexander Ludwig, new to the team and yet to be sent out on a mission, must literally perform for the rest of the men. He does a solo dance to Jamiroquai's 1999 dance hit "Canned Heat" and then recites a poem. The dance is framed as a joke: the homoerotic implications of the moment suppressed by individual SEALs covering their eyes and the group quickly calling a halt to the performance. But the poem is presented as deadly serious. This part counts: all the main characters ask, "Can you say it?" The young SEAL recites "Ballad of the Frogman." The poem—used as a toast; its original author unknown—can be found in multiple versions on different SEAL and military community websites. *Lone Survivor* offers an expurgated version. Pointedly lacking are the lines, common to most versions, "I dive for five / Tuck, suck, fuck, nibble, and chew / I dine and intertwine, masturbate, ejaculate, and copulate," as well as the most common concluding words, "and ladies, if you don't like my face you can sit on it!" The film includes more moderate sexual euphemisms, like "no muff too tough," and inserts generic inspirational sentiments out of tone with the earthy original, like "Anything in life worth doing, is worth overdoing; moderation is for cowards." Whose sensibilities are being preserved here? I saw *Lone Survivor*

at a matinee the weekend it opened in suburban south St. Louis, a predominantly white area of the city. The theater was packed with an audience representing all four quads. Censoring "Ballad of the Frogman," especially given
its structural placement at a hinge point in the narrative, likely was deemed
prudent for the older women anticipated to be in the audience (and indeed
there in numbers that afternoon). *Lone Survivor* is not just for men: like
Saving Private Ryan, its tone is calibrated for mixed company and multigenerational comfort. Of course, anyone watching *Lone Survivor* who knows the
unexpurgated "Ballad of the Frogman" can revel in their secret knowledge of
what the SEALs are really like.

Another military creed has had a recurring place in the PCF, linking
films to each other and to military tradition. "The Rifleman's Creed," written by Marine Corps Major General W.H. Rupertus in the early 1940s,
remains part of Marine recruit training. Beginning with the line "This is
my rifle," the poem shows up in several films. Each inflects the text differently. *The Boys in Company C* only nods toward the "Creed," with the
fragmentary line "without my rifle I am useless" signaling those in the
know. *Full Metal Jacket* incorporates the "Creed" in full. (Hasford's novel
includes the complete text.) Drill Instructor Hartman has the men climb
into their bunks as if on the parade ground with rifle in hand and recite the
"Creed" together as pillow talk to their rifles. To ensure that the men spoke
in perfect unison, Kubrick had Matthew Modine, playing Joker, record the
words "to a click track—a metronome cadence used," in Modine's words,
"to keep us all in sync."[30] Then, while shooting the sequence, each man in
the shot listened to and recited with the recording by way of a wireless
earpiece. Here, Kubrick exercises a measured rhythmic control over soldiers' collective speech by invisible technical means. (Colby Buzzell tells in
My War [2005] of writing his own variant of the "Creed" after seeing *Full
Metal Jacket* and taping it to his bedroom wall, an example of how PCFs
prepare young men for war and military culture.) *Jarhead* includes a call-
and-response version of the "Creed," overlaid with Swofford's voice-over
about his desire for the "pink mist," the head shot, the chance to kill with
skill. Later Swofford repeats the "Creed" at a moment of extreme tension,
when—out of frustration, boredom, and anger at himself—he points his
rifle at a fellow Marine who fell asleep while taking Swofford's watch on
Christmas Eve. The measured cadence of the first recitation of the "Creed"
is lost in the shouts, tears, and desperation of the second. Self-control—one
key to American soldiering—is almost lost. *Jarhead*'s edgy exploration of
the soldier's identity, his nearness to losing it, is thrown into relief by the
film's dual use of the "Creed."

A third example of authentic military language used in the PCF starkly links military training to sexually aggressive masculinity. In the book *Flags of Our Fathers*, James Bradley describes the process of training young citizen soldiers, many of whom had no knowledge of firearms. He writes, "Above all, the rifle was a *rifle*. Make the dipshit mistake of calling it a 'gun' and you got humiliated before the entire company—forced to run up and down in front of your buddies in your skivvies, one hand holding your rifle and the other your gonads, screaming over and over: 'This is my rifle and this is my gun! One is for business, the other for fun!'"[31] Kubrick, drawing on another episode in Hasford's novel, turned this practice into a musical number, with the entire group of Marine recruits marching through their squad bay in time to the phrase.[32] Hartman, in full uniform, leads the charge, with the men following in two single-file lines on either side. The men are stripped to T-shirts and boxers. Kubrick creates a completely symmetrical picture by putting rifles in each line's outside arm. No conventional military drill would produce half of a group shouldering arms right and half shouldering left. The men's free inside hands grasp their crotches, giving a big squeeze to their genitals in rhythm with the words "*this* is my *gun*" and "*this* is for *fun*." The position of their weapons and the addition of a rhythmic hand movement makes this nothing more or less than a dance number, with Hartman as choreographer and star. Kubrick takes an authentic military practice and stylizes it by musicalizing it for satiric effect—comic or horrific, depending on the viewer's perspective.

In *Casualties of War*, just before he moves to begin the rape of Oanh, Meserve distinguishes between his weapon and his genitals by way of the familiar phrase. Meserve confronts Eriksson, concluding that he must be gay if he doesn't want to "take his turn" with the girl, then threatening, "Maybe when I'm done with her, I'm gonna come after you. Maybe when I'm done humpin' her, I'm gonna come hump you." On this, Eriksson pulls his rifle on Meserve, who challenges Eriksson's "attack posture" by throwing his own rifle to him. Meserve notes that everyone has a weapon, everyone can blow anyone away at any time. Done with talking, Meserve approaches Eriksson to retrieve his rifle and says, "The army calls this a weapon. But it ain't." Moving toward the hooch where Oanh is tied up and grabbing his own crotch, Meserve says, "This is a weapon." He refers to the rifle in his hand with the words, "This is a gun." Shaking his genitals he says, "This is for fighting." Shaking the gun, "This is for fun." The rape sequence follows, a series of cross-fades of each man taking his turn with Oanh while Eriksson sits some distance away. Meserve's perverse reversal of the phrase verbally constructs the moment as outside the disciplined environment of the military. As Eriksson says to Meserve after the rapes have occurred, "This ain't the army."

The Vietnam War was particularly rich in soldier slang. Most films in the 1980s PCF cycle draw on this lingo, a strategy that ends up tying these films to the realities of soldier life (and the literature of the war) and linking the films to one another. This double situation makes it difficult to know if a film from later in the cycle is quoting an earlier film or simply tapping the same rich vein of authentic soldier talk. For example, *The Boys in Company C* tosses in a fragmentary reference to a common soldier adaptation of Psalm 23: "Yea tho' I walk through the valley of death, I will fear no evil. 'Cause I'm the baddest motherfucker in the valley." Often engraved by American soldiers in Vietnam on their personalized Zippo lighters, Hasford recommended the saying to Kubrick for use as graffiti on helmets or flak jackets in *Full Metal Jacket,* and the ever-menacing Meserve declaims the full text in *Casualties of War.*[33]

The phrase "sorry 'bout that" turns up regularly in the literatures of Vietnam: in oral histories such as *Headhunters: Stories from the 1st Squadron, 9th Cavalry in Vietnam, 1965–1971* (1987); in literary memoirs like Philip Caputo's *A Rumor of War* (1977); and in novels like Daniel Ford's *Incident at Muc Wa* (1967, source for *Go Tell the Spartans* [1978]), where the phrase is spoken seven times, each time by a different soldier at a different rank, always in response to a shitty situation.[34] (Curiously, "sorry 'bout that" appears three times, in prominent moments, in the published script for *Platoon* but never in the film itself; Stone, well aware of how soldiers talked in Vietnam, seems to have decided against larding the dialogue with the phrase.) The dismissive imperative sentence "Sorry 'bout that" takes on a central role in *Casualties of War.* Just before he's shot, Brown is teaching Eriksson—a cherry who has been in country three weeks—when to use the phrase and how to say it. The goal is to harden Eriksson toward the Vietnamese people. The young man never quite gets it right. Indeed, Eriksson never learns to say "sorry 'bout that," but instead clings to a simple language of personal moral responsibility. He says "I'm sorry" two times to Oanh's mother as they kidnap her daughter. He ends his recounting of Oanh's rape and murder to a Methodist chaplain with a simply stated self-incriminating sentence: "And I failed, sir, to stop them." And in the court martial scene—cut in the film, available in the extended-cut DVD—Eriksson says of his actions, "I did nothing." All these "I" sentences give the lie to the dismissive, responsibility-shirking, passive-aggressive imperative refrain "Sorry 'bout that." Eriksson's inability to *not* speak of himself as responsible captures in the domain of grammar the film's core critique of any US cultural memory of the war that fails to claim responsibility for the damages inflicted on the people of Vietnam.

There is a similar moment in *Platoon*. No line gets to the heart of the 1980s PCF cycle like Chris's words after he stops men from his platoon from raping a young girl at the climax of the village atrocity scene. One among the group says, "What the fuck's your problem Taylor. It's just a fucking dink." To which Chris replies, with tremendous passion, "She's a fucking human being, man." This line appears nowhere in the script record for *Platoon*. It was likely added during the shoot. Its directness—cutting to a fundamental moral issue of the war—is but one mark of the complex moral goals of *Platoon* as both action film and meditation on American guilt and innocence. Spoken in the midst of a fully rendered re-creation of the war, including thrilling combat sequences, Chris's added line marks a high point in PCF (and Hollywood) history—a moment when a commercial film pushes the viewer to reflect, the film's maker unafraid not to entertain, eager to arouse moral passion, and here to challenge soldierly behavior and the conduct of the war as a whole. Chris's line is expressly political. A major strand of antiwar critique is distilled in his insistence that the Vietnamese girl—a foreign other—is indeed a "human being."

A final Vietnam motto: "It don't mean nothin'." The veteran and military critic Andrew Bacevich writes, "By Vietnam 1971, the army's unofficial motto had . . . become 'don't mean nothin',' usually muttered sotto voce at the back of some annoying superior."[35] The phrase, also often engraved on Zippo lighters, is memorably used in several films. King, the wise black soldier who finishes his one-year tour and heads for home just before the final battle in *Platoon*, uses it prosaically in his final talk with Chris as a last bit of wisdom from one who survived. In *Casualties of War*, Meserve says "It ain't nothin'" to the mortally wounded Brown just before he's heloed out. Brown replies, "It sure feels like something, man." But no film exploits the flexible resource of "don't mean nothin'" like *Hamburger Hill*. Critic Jack Kroll singled out the phrase and its relationship to race, a major topic of the film: "The tension between white and black grunts is resolved in their mutual acceptance of the blacks' rhythmic motto: 'Don't mean nothing, not a thing.' This Motown metaphysics sums up the spiritual dislocation of a war whose meaning continues to mock us with its lethal ambiguities."[36] While the phrase has nothing to do with Motown, Kroll rightly highlights its rhythmic use and its connection to black physical performance. The *Hamburger Hill* press kit synopsis describes the scene just after the death of the black grunt McDaniel: "He is mourned in a special way when WASHBURN joins DOC and MOTOWN as they ritualistically DAP each other. Some people might say it's just three niggers shaking hands; a lot they know." Some critics—like some viewers, no doubt—didn't recognize the dap: one called it "patty-caking and chanting."[37] The dap was a physical greeting shared by black servicemen in Vietnam: a sign of solidarity and mutual care. In the film, a sequence of handshake movements is combined with rhythmic repetition of the phrase, "It don't mean nothin' / Not a thing." As the black trio in the squad move and speak together in a blues-like moment of grief transformed into joy, the white soldiers around them watch, some mouthing along, trying as they can to join a moment constructed as racially exclusive. Later, *Hamburger Hill* uses an interracial handshake to make the film's central point, when racial tensions within the unit are resolved by the smallest possible dap, a single fist bump between the black medic Doc and the white working-class ethnic Beletsky, the former hailing the latter as "brother blood." Doc puts one moral of the film into words when he exits *Hamburger Hill*, saying, "We're all niggers on this hill." Carabatsos may be idealizing the dap as a means to racial reconciliation. News stories quoted by Andrew J. Huebner emphasize how black soldiers used the dap to irritate or subtly threaten the white soldiers excluded from the greeting.[38] *Hamburger Hill*, while presenting the dap as important for black soldiers in Vietnam, frames the rhythmic and physical practice as something whites envy rather than fear and something blacks are prepared, on some level, to share. Vietnam veteran Karl Marlantes includes a scene of instruction in the dap between a black and a white soldier in his 2010 novel *Matterhorn*. The white soldier learns the moves but says, "It just doesn't feel right." The black soldier replies, "It never will. . . . You ain't black."[39] Here, the dap fails to bridge a persistent racial gap. As shown in the next chapter, sharing black music proves a sturdier means to cross-racial understanding.[40]

A final PCF refrain, heard across the Gulf War and GWOT cycles, brings this chapter to a close. In the opening scene of *Courage Under Fire* (1996), Lieutenant Colonel Serling caps off his prayer before battle with the words "Kill them all." His men repeat the phrase, which turns tragically ironic after Serling kills his own men in an act of friendly fire that haunts him to the final reel. "Kill them all" concludes *The Kingdom* (2007), when it is revealed as both the words a dying terrorist bomb maker whispers to his young grandson and the phrase an FBI agent whispers to a coworker to comfort her after the death of a colleague in a terrorist bombing. Finally, a SEAL in *Lone Survivor* announces that he intends to ascend the mountain, rescue his comrade, and "kill them all" (the attacking Taliban) at a moment when doing any of these things is literally impossible. There's ample evidence real soldiers use the phrase. In the book *Lone Survivor* (2007), Marcus Luttrell has his buddy Mike Murphy say "Kill them all" during the heat of the battle in which Murphy lost his life.[41] David Finkel lists "Kill them all" among responses to "a game of what [soldiers] wanted their last words to be." (Other responses included "Fuck Nine-eleven" and "Tell my wife I really didn't love her.")[42] In the PCF, "kill them all" is never evoked in uncomplicated ways but instead resounds as ironic, as misguided, as delusional, as a sentiment linking American heroes and their enemies, as a knee-jerk reaction by military men (and others in the business of killing) that the film viewer is invited, should they wish, to question. It's never deployed as a self-satisfied one-liner.

4.   Soldiers' Song

Singing is an act of soldierly agency. Sometimes it briefly takes a soldier out of the war: Private Mellish in *Saving Private Ryan* (1998) sings a lonely phrase of the Duke Ellington song "Solitude" while walking into German-occupied France. Other times, a soldier sings to revel in his power over others: Clark in *Casualties of War* (1989) walks beside Oanh before the squad has raped her and creepily croons "Hello . . . hello . . . . Hello, I love you won't you tell me your name." The camera responds by tilting sharply upward to the sky, as if in revulsion or shame. A viewer familiar with the Doors' 1968 record "Hello, I Love You" might be forgiven for never thinking of that song in the same way again. (Clark also "woos" Oanh with a phrase of the Dean Martin hit "Everybody Loves Somebody Sometime," but the effect is likely not the same for viewers who came of age after the 1960s.) Other times, the urge to sing is physical. Ray Person, Humvee driver in *Generation Kill* (2008), drives the other Marines in his vehicle crazy with his constant singing—except when they join in like a family on a road trip. In one scene, Person awakens from a nap and asks, in a hurt voice, "You guys sing 'King of the Road' without me?" At the end of the series, Person falls quiet: his constant singing and chatter revealed as a side effect of the Rip Fuel he drank in quantity to stay awake through the invasion.

The 1943 film *Guadalcanal Diary* is so full of singing soldiers that the combat film scholar Jeanine Basinger dubs it the "combat film Hit Parade." Basinger argues that singing works to mythologize these guys.[1] Could it be that American men—specifically soldiers—just used to sing more? And what do they sing? Familiar songs that bind this white man's army into one. The film's first song is the Protestant Christian hymn "Rock of Ages." Its first joke comes from a Jewish soldier singing along: "Why not, my father was a cantor in the synagogue," he says. Old-fashioned favorites like

"I Want a Girl (Just Like the Girl That Married Dear Old Dad)" are heard in full harmony. Harmonica echoes around the ship: playing "Chattanooga Choo Choo" (for two men dancing a wickedly athletic jitterbug), an aloha song (to which a fellow performs the hula), an Irish jig (in honor of the ship's Catholic priest, predictably from Notre Dame), and "My Old Kentucky Home" (a quavery solo while the men read letters from home). One night before battle, the men write letters home—heard as voice-overs—to the sung strains of "Home on the Range." Many of these musical moments come directly from the 1943 book *Guadalcanal Diary*. Author Richard Tregaskis describes how the men sing while they work, make up rhythmic chants about killing "Japs" while loading machine-gun belts, and set new words to popular songs. Marines are heard singing in the night songs like "Blues in the Night"; with a lyric about a "two-faced" woman, to Tregaskis it "seems to be the most popular song."[2] (Soldiers dwelling on possibly unfaithful wives and girlfriends is a deep theme in combat stories: it dominates *Jarhead* [2005], for example.)

Guadalcanal Diary offers access to past musical practices among soldiers that, for their original audiences, were expressions of the shared present. PCFs similarly capture soldiers engaged with popular music—singing, dancing, listening, and also talking about music as part of their shared lives. Generational differences between the men who fought World War II, Vietnam, and various American wars in the Middle East account for some of the variety in the use of popular songs in respective PCF cycles. Popular music and dance styles and technologies change, shaping each generation into different musical cohorts. But in the PCF, popular music also plays a role in relation to the film audience: some past musics have greater power for PCF audiences than others. And, as always, there's the issue of contemporaneity in the case of the GWOT cycle in which, as in *Guadalcanal Diary*, music and dance potentially work to define the present. This chapter looks at the use of popular music in the PCF. The emphasis throughout is on music that is part of the world of the film: diegetic music either sung or listened to by soldiers. Such moments, and there are many, bring not only music but also dance and the linguistic content of lyrics, which often eloquently enhance the verbal worlds of these films. Popular music as a topic of conversation and contention also enters and amplifies the stories these films tell about American soldiers abroad.

A popular music divide between pre– and post–rock and roll wars is evident across the subgenre. World War II PCFs, set before rock and roll, have virtually no popular music. In PCFs depicting Vietnam and later conflicts, soldiers sing and listen to rock and roll, country, pop, heavy metal, and hip-hop that is charged for movie audiences with the social transformations of the 1960s and after. In many cases, filmmakers use diegetic popular music to highlight the class, racial, regional, and political identities of individual soldiers or groups. Often references to song titles and lyrics appear in early scripts, revealing popular music as a constituent element of the conception of a given scene, character, or film (even if the named song did not make the finished film). The changing nature of the military also shapes the use of popular music: the citizen soldiers of World War II, the draftee army of Vietnam, and the all-volunteer force of the post-Vietnam era, including the elite warriors of the GWOT, each enjoy music differently and to different thematic ends for film-makers. As with any aesthetic element, the storytelling and character-building resource of past or present popular musics serves filmmakers making a film for commercial release in a specific context. Use of compiled cues entails balancing competing priorities, including the financial and industrial considerations that always shape the presence of popular music in film.

JUMPING CULTURAL GAPS IN POST-ROCK PCFS

The sonically spare *84 Charlie MoPic* (1989) offers a benchmark for music in the Vietnam cycle. The squad's friendly radiotelephone officer (RTO) is a talkative fellow who appreciates his role as the group's connection to the outside world. At one point, during a rest break, he picks up AFVN (Armed Forces Vietnam Network) on his radio. Everyone, including the standoffish black sergeant, listens thoughtfully to the music: a minor hit from 1965 by the British singer-songwriter Donovan called "Catch the Wind." The musical moment is a respite for the viewer as well as for the men, who are each shown in turn listening intently. Of course, this music break is an absurd lapse of noise discipline, advertising the squad's position in the jungle to any enemies nearby—and in Vietnam the enemy is always nearby. The RTO looks apologetic when he loses the signal. The broadcast, heard at a low volume through a tinny speaker, presents recorded popular music as an important part of this war without putting much in the way of specific musical content into the point. It does not elicit audience response by way of prior familiarity with a genuine hit, nor does Donovan's folky sound carry especially divisive race or class implications. None of the men respond to its style or genre. Here, popular music offers simple comfort. Most Vietnam films use better-known music with greater specificity in terms of musical genre and style to efficiently make precise points.

Popular music in Vietnam PCFs typically marks important divisions of race and class among the men and activates generational knowledge in the film audience. In part, this opportunity comes from the musical advantage Vietnam enjoys over other conflicts. Popular music of the 1960s, in all its varieties, remains a potent cultural force, especially for baby boomers but also for subsequent generations. Gen X soldiers, like Swofford in *Jarhead*, know Vietnam-era music. When an unseen helo flies by in the night blasting a track by the Doors, Swofford complains, "That's Vietnam music, man. Can't we get our own fucking music?" The nostalgic power and market muscle of 1960s music for Hollywood audiences was activated by hit films like *American Graffiti* (1973) and *The Big Chill* (1983). The use of vintage popular music in Vietnam films should be understood within this larger trend. For example, the PCF-adjacent comedy-drama *Good Morning, Vietnam* (1987), about an AFVN disc jockey, is built around popular hits of the era played in a story-driven context. The commercial power of this old music proved itself yet again. Louis Armstrong's 1967 record "What a Wonderful World," used in *Good Morning, Vietnam* behind a montage of American soldiers at work and play, was re-released as a single in conjunction with the film and charted in the Top 40.

On a thematic level, 1960s popular music proved a ready resource for characterization of racial and regional divides. In this, diegetic pop music in the Vietnam PCF cycle reflected the experience of veterans. The American presence in Vietnam was awash with the sound of popular music. In his 1991 Vietnam memoir *Acceptable Loss: An Infantry Soldier's Perspective*, Kregg P. J. Jorgenson describes GI bars on the same street catering to different clientele, playing "soul music" or "country-western music," while in another "Mick Jagger was screaming that he couldn't get no satisfaction." Jorgenson continues, "A tape deck provided music and GIs returning from R&R in Hong Kong, Taiwan, Australia, or Thailand brought back reels of their favorite music, which ran the spectrum from country and western to soul. Cultural gaps were jumped musically while the soldiers drank canned beer or war whiskey."[3] J. Martin Daughtry notes how "portable cassette players," widely available during the Vietnam War, "marked a pivotal moment in the history of mobile music and war, a moment in which a new technology facilitated a democratization of listening practices in theater."[4] This technological innovation proved ideal for film, as individual characters could be responsible for specific musical choices. Playing a cassette tape—as Clean does, for example, to Willard's annoyance on the patrol boat (PBR) in *Apocalypse Now* (1979)—becomes an act of musical agency akin to singing. Such musical acts opened cultural gaps capable of causing conflict within the ranks. One Vietnam vet described how racial tensions erupted over musical differences: "One battalion I know of came in for a five day stand down and ended up being sent back out the same day they arrived because they tore the EM club apart over [whether to play country or soul music]".[5]

Music marking racial, class, and ideological identities, causing conflict, and bridging "cultural gaps" is a recurrent trope in the Vietnam cycle. The lone black soldier in *Go Tell the Spartans* (1978), a radio operator nicknamed Toffee, is briefly characterized by the sophisticated jazz he listens to on the job. Ron Kovic's brother in *Born on the Fourth of July* (1989) is heard learning to play Bob Dylan's "The Times They Are A-Changing," telegraphing his later resistance to the war. Blacks and whites in *The Boys in Company C* (1978) bond over a guitar and harmonica duet performed in the back of a six-by truck: a white soldier plays; a black soldier sings the blues. The latter says to the former, "You must have some black in you." A voice-over describes the moment—and the pee break after—as "the pause that refreshes" (quoting an advertising tagline for Coca-Cola in a further pop-culture reference).

Hamburger Hill (1987) develops issues of musical taste at length, using music to move toward a reconciliation of racial tensions within the group, marked by Doc's line, "We're all niggers on this hill." The film generally opts for the discussion of music rather than performance per se, although actual performance was part of writer Jim Carabatsos's conception. The script frames three of the four black men in the troop as a kind of musical group: one is named Motown (the name survives in the film). The original script and press kit synopsis include a scene not in the film featuring the black trio performing together for Sergeant Frantz "a precision step while they *mouth* the words to [Smokey Robinson's "I Second That Emotion"] on Motown's CASSETTE PLAYER."[6] Frantz as a sympathetic, racially enlightened white officer is balanced by Sergeant First Class Worcester, who, in a line that made the film, complains to Frantz in his Southern accent about AFVN radio—"Don't they ever play Tammy Wynette?" Motown offers a complementary complaint about country music, also to the mediating figure of Frantz, "I don't want to hear any more of that rebel 'I lost my car on the motherfucking road and I'm crying over you' shit" and laments the loss of an FNG (fucking new guy) who "knew everything the Temptations ever did." The death of McDaniel, one of the dancing trio of black men, is dealt with in the script by musical means. The remaining two "make a half-hearted attempt at a dance routine" to "I Heard It Through the Grapevine." When the fourth black soldier in the troop—Washburn, the lone black cherry—steps up and adds a new step, called the "Point," the reconsolidated trio bursts into a dance to Aretha Franklin's "Since You've Been Gone." Doc, leader of the group, declares, "The 'Brothers' are back." Carabatsos notes in the script: "THE BROTHERS dance in unison as they mouth the words to the song. Some people might say it's just three niggers dancing; a lot they know." As described in the previous chapter, in the film the trio reacts to McDaniel's death by performing a dap to the words, "It don't mean nothin' / Not a thing." Perhaps the dance number just didn't work in practice.

Oliver Stone concisely captures racial "gaps," and the power of black music to bridge them, early on in *Platoon* (1986). A sequence of three songs—arranged in an ABA pattern—characterizes the platoon's two contrasting groups and sergeants in musical terms. Jefferson Airplane's "White Rabbit" plays for the introduction of Chris—Stone's autobiographical "white Infantry boy" stand-in—to marijuana in the bunker with the "heads," including Sergeant Elias. A direct cut to the barracks and a poker game, including Sergeant Barnes, has Merle Haggard's "Okie from Muskogee" playing on the radio. The song's opening line, "We don't smoke marijuana," puts song lyrics to work delineating different groups of soldiers by race and class. In the barracks, the black soldier Junior derides Haggard's song as "redneck noise" and "honky shit," and says he wants to hear some Motown. Stone concludes this reflective section of the narrative by cutting back to the "heads" for a group dance to Smokey Robinson's "Tracks of My Tears." The tune fades out into the start of a long, unscored jungle patrol sequence. Nothing substantive happens during the Robinson number except that Chris, Elias, and others—significantly the black men in the platoon—bond over music, dancing with each other in the red light of their underground refuge. When I asked Oliver Stone about the music in *Platoon* at a conference devoted to his films at Rider University in 2013, he candidly volunteered that "dope, music, and a little love between men" sustained him during his time in Vietnam. (Stone also said that he had not heard black popular music until arriving in Vietnam; cut lines from the scripts for *Platoon* and *Born on the Fourth of July* both include a character eager to introduce someone to the music of Jimi Hendrix.) The ABA popular music sequence in *Platoon* makes this evident: the A songs express the pleasures of dope, provide a suitably soulful sonic setting for "love between men," and solidify for the viewer which group the film is surely on the side of; the B song interrupts the party spirit, equates country music with the violent side of the platoon (a perhaps predictable class and region-based argument from the Northeastern-born Stone), and makes more precious by comparison the heads' hideaway. The silencing of "Tracks of My Tears" on the fade into the jungle patrol underlines the film's shift from reflection—a pop music ABA using nostalgic, memorable 1960s songs—to action.

Stone's use of "White Rabbit" rewards close analysis. An early script draft spots "Tracks of My Tears" as the music for this scene and includes a long, crude stretch of dialogue for the heads. Elias comes in "fresh from the shower." Hoyt and Big Harold pull off his shorts, leading to a wrestling melee for the three, "like kids." When Chris takes his first hit, Manny, Big Harold, Francis, and Doc join in on "a high falsetto snatch of blues." The scene ends with a "Saturday night dance party. A yearning for tenderness, for femininity, for a moment of peace in this nightmare life."[7] A later script eliminates almost all the dialogue and the wrestling melee, lists "White Rabbit" as the song on the tape deck, and introduces the memorable action of Elias "shotgunning" Chris, expelling a mouthful of smoke into the new guy's mouth through the barrel of a rifle.[8] In the film, events in the bunker closely follow the musical course of "White Rabbit," creating a song scene—where a preexisting piece of recorded music heard in its entirety provides a template for the unfolding of events.[9] The song literally shapes the scene. The signature bass riff opening "White Rabbit" comes into the soundtrack as scoring at the end of the previous scene, as Chris tokes up against the setting sun. As Chris enters the bunker, the sound of the track is gently futzed, transitioning the recording into diegetic status. Instead of the men talking, as he originally imagined in the script, Stone offers instead images of the men listening to music. The length of the track opens up a lyrical interlude where the physicality of the men can be highlighted. The image of Manny—smiling, facing the camera head on while doing barbell curls—is re-seen in the visual roll call at the end. It's hard to imagine a context for such an image in a nonmusical scene: with "White Rabbit" sounding in the space of the bunker, there's time to simply look at Manny (and the others) being themselves. "White Rabbit" plays from start to finish. Stone exploits the musical form of the track, its long, gradual crescendo, building to the "shotgunning" climax as singer Grace Slick wails "feed your head" three times. Just after the song ends, Chris coughs, the sound, in the words of Stone's script, "kicking off the next image" on a hard cut to the barracks tent, where the opening lyric from "Okie from Muskogee" offers its direct response.[10] Replacing dialogue with music and using the lyrics and musical form of an interpolated pop song, Stone's song scene to "White Rabbit" powerfully incorporates soldiers listening to pop music into the warp and woof of military life.

Popular music proves a ready resource for characterization of racial and regional divides in the Vietnam cycle. In these films, popular music marks divisions, and white characters who are able to cross those lines—Chris and Frantz—are framed as enlightened. In the Gulf War films *Three Kings*

(1999) and *Jarhead*, shared musical tastes among black, white, and Latino soldiers present the all-volunteer, post-hip-hop military as a zone of musical and racial tolerance. Musical differences become matters for humorous back and forth: not, as in *Hamburger Hill*, cause for soldiers to fight among themselves in earnest.

After a short opening sequence, *Three Kings* explodes with "frat-party abandon," the Americans victorious in their effort to remove the army of Saddam Hussein from Kuwait.[11] A medley of four diegetic tunes opens the film:

- Rare Earth's classic rock track "I Just Want to Celebrate" (1971) presents the Gulf War as a beach party, with visual references to *Platoon* (Manny's barbell curls) and *The Big Lebowski* (1998; a shirtless soldier uses a huge water bladder as a trampoline). The exuberance and cocky bravado of the soldiers is palpable even though the film has not shown them proving their mettle in combat.

- Next, Lee Greenwood's patriotic country-pop anthem "God Bless the USA" (1984) is sung into the film by a series of singing soldiers, eventually arriving at the tent where the main characters and a racially diverse group of soldiers shout along with the record. These guys can't sing, but it doesn't matter—or perhaps that's the point.

- Public Enemy's "Can't Do Nuttin' for Ya Man!" (1990) cuts in next, spun by a white soldier DJ, setting military victory as locker room champagne party. Actor Mark Wahlberg, as soldier Troy Barlow, leads the dancing; Wahlberg's moves draw a connection to his recent film *Boogie Nights* (1997). The track's thumping beat cuts from the men dancing with each other to Archie Gates, played by George Clooney, having sex with a reporter. A short scene introducing the topic of the missing Kuwaiti gold follows before an abrupt cut back to the party, where the men are grooving to . . .

- Snap!'s 1990 dance hit "The Power." The scene has grown playfully sexual—Troy and Conrad, a Southern white soldier who idolizes Troy, exchange haircuts; Troy calls Conrad "very handsome"—and playfully violent, climaxing with a series of screams in each other's faces. In one shot, Conrad screams and holds a pistol in his hand. The beat shuts off on a direct cut to the next morning.

Three Kings marks "victory" in the Gulf War with an all-male party touching on several musical eras and styles. There is no conflict over music here: everyone knows the words and grooves to the same playlist. Music talk and trash talk, much of it on racial topics, unfolds across the film but never with any real stakes. Conrad, the "cracker," is belittled as ignorant and an embarrassment. Still, he's family—although by tacit understanding excluded from the film's hip and handsome, interracial title trio. Conrad's death proves a genuinely sad moment.

Other uses of diegetic music and talk about music develop *Three Kings*'s theme of pop-culture connections between the Americans and the Iraqis. When Troy is being interrogated—and briefly tortured—his Iraqi interrogator asks, "What is the problem with Michael Jackson?" and calls Jackson the "pop king of sick fucking country." Briefly encountering an Iraqi man listening to Eddie Murphy's "Party All the Time," Troy turns it off and advises him in earnest tones, "That's bad music. This music is bad for you." (In one script the "bad music" is Olivia Newton-John's "Physical.")[12]

Jarhead similarly suggests an early-1990s pop-music consensus among the all-volunteer force. A drunken Christmas party has black, white, and Latino men dancing and rapping in a circle to "O.P.P.," the rap group Naughty by Nature's crossover hit from 1991—although the track was released some seven months after the Gulf War concluded, making the scene technically impossible. (Non-diegetic pop tracks in *Jarhead* include three big hits of the film's time frame: pop [Bobby McFerrin's "Don't Worry Be Happy"], dance [C&C Music Factory's "Everybody Dance Now"], and grunge [Nirvana's "Something in the Way"].) *Jarhead* concludes with a bonfire bacchanal including celebratory if frustrated firing on automatic into the night sky to the strains of Public Enemy's "Fight the Power," a more contentious rap song that was not a pop hit. One script draft called for "thrash metal" at the closing spot.[13] In the end, either black or white music of noise and power would apparently do for the makers of the film. Country—always the contentious white genre—is only suggested in *Jarhead* by the novelty song "Grandma Got Run Over by a Reindeer," which is tied to one white soldier and not the whole group.

ROCK AND THE AMERICAN WAY OF WAR IN THE DRAFT SCRIPTS FOR *APOCALYPSE NOW*

> "Certain rock and roll would come in mixed with rapid fire and men screaming."
>
> —MICHAEL HERR, *Dispatches* (1977)

Like Stone and Carabatsos, *Apocalypse Now* screenwriter John Milius and writer-director Francis Ford Coppola could not imagine the Vietnam War without soldiers listening to and talking about popular music. (A tantalizing cut scene included Kilgore at the beach party the night before the helo attack playing a guitar, his "SWEET, MELODIOUS VOICE SINGING a very sad song" that the dialogue reveals he also wrote.[14] In the film, Kilgore briefly strums a guitar.) Coppola included the following sonically evocative description in a 1975 script: "The RADIO is still left on, putting out STATIC, strange fragments of MUSIC, as though the air waves of Vietnam are an amalgamation of our rock music, our calls for air strikes and a thousand technical conversations."[15] This thick sonic texture echoes Milius's description in his original 1969 script draft of the tape of Kurtz in the jungle played in an early scene: "A heavy BURST of AUTOMATIC WEAPONS FIRE—INSANE LAUGHTER—STATIC, and faintly, very faintly we HEAR HARD ROCK MUSIC more STATIC—suddenly a low, clear VOICE [Kurtz] peaceful and serene, almost tasting the words."[16] "HARD ROCK MUSIC" was a guiding sonic reference point for Milius and Coppola alike as they crafted the film.

Milius named specific tracks in his original script, and Coppola added to and altered the film's playlist in subsequent drafts. Some of these songs ended up in the film. Creedence Clearwater Revival's 1968 hit "Suzie Q" was Milius's first choice for the USO show featuring Playboy bunnies. In the film, it's performed by the band on the floating stage. The Rolling Stones' "(I Can't Get No) Satisfaction" on Clean's cassette player accompanies Lance waterskiing behind the PBR. Coppola added this scene and described the music and the men's reaction to it: "The SONG BLARES ON—they all dig it."[17] The game of chicken between the pilot Chief's and another PBR was originally scripted as initiated by music. Chief's question "Is that you, Lazzaro?" was to be answered by "a short BURST of ROCK MUSIC over [the approaching PBR's] speaker." As the boat drew close, "Suddenly the ROCK MUSIC BLARES again—some Grateful Dead—as it swerves at the last minute and fishtails past. One of it's [*sic*] crewmen moons our crew."[18] Again and again, Coppola's soldiers listen and respond to rock music, mostly from the late 1960s and mostly pinpointed down to the track. Coppola's Vietnam is awash in late-1960s rock—not the Motown prevalent in *Platoon* and *Hamburger Hill* or the folky, acoustic guitar touched on in *The Boys in Company C* and *84 Charlie MoPic*. *Apocalypse Now* is a rock war film. The opening sequence set to the Doors' "The End" makes this point succinctly, if in the zone of Willard's head.

But the use and resonance of rock as a music of chaos and destruction was developed in much greater detail in Milius's and Coppola's imaginations than in the finished film. Indeed, the major difference between the scripts and the film as made is the elimination of combat sequences set to diegetic rock. Kilgore's helicopter attack, the film's only set-piece battle, is, of course, fought to diegetic music. In essential ways, Richard Wagner's "Ride of the Valkyries" was treated by the film's makers as a rock record in terms of both the use of amplification and the scaffolding of film form on the musical form of a specific recording (see chapter 7). The scripted-but-never-shot scenes described below provide a context for Kilgore's musical helicopter attack. These scenes underline how rock music in *Apocalypse Now* was approached not as nostalgia but as musical expression of the fundamentally destructive nature of the war as prosecuted by the United States. Milius and Coppola identify late-1960s rock as a musical analogue and perhaps even enabling aspect of the perverse acts of violence enacted by the American officers and soldiers portrayed in the film. Rock expresses the war's essential nature, just as the opening words of the film and the opening lyric of "The End"—"This is the end"—restate the film's title.

Milius's 1969 script opens deep in the jungle at Kurtz's compound. His American and Montagnard forces are bizarrely attired in loincloths with bandoliers of ammunition strung across their chests. Kurtz's first words—the script's first line of dialogue—is the quasi-Shakespearean command: "Music—Sergeant blare." As shown above, the verb *blare* is a favorite in the scripts. The word speaks to amplification as an essential quality to war music in Vietnam. Kurtz's command is immediately answered: "Massive loudspeakers out in the middle of a song—the Cream [*sic*] singing—'Sunshine of Your Love' in all its blazing electric harshness. The men pass through the gate—proud defiant."[19] Kurtz's soldiers march into battle to a rock beat. The ambush of a North Vietnamese Army (NVA) unit follows, opening the film with specific, ground-level combat rather than the stylized images of napalm air strikes of the finished film. The lyric for "Sunshine of Your Love" describes a consummation with a lover in the darkness of night for which the singer "has been waiting so long." Heard as a jungle battle song, the lyric equates combat with an encounter with primitive violent forces, including death itself. In his 1975 script, Coppola added a musical close to Milius's opening battle. One of Kurtz's men "opens his flame thrower directly ON US and the NVA soldier and we are incinerated in flame, bright psychedelic orange-red flame. Outrageous, loud, electric ROCK MUSIC OVERWHELMS the SOUNDTRACK." The title sequence was to follow, using the visual motif of fire "growing more intense, brights [*sic*], more vivid, purifying; transforming into an intense white heat that we can barely look at, like the sun itself. Then it EXPLODES."[20] Compared to these scripts, the film's opening montage set to "The End" seems positively sedate.

In the original combat-centered close to the script, the Doors' 1967 track "Light My Fire" was to serve as battle music for Kurtz's army. The specificity with which this record is woven into the scripts proves striking. Coppola's final draft had Kurtz's natives singing along with the Doors' lead singer, Jim Morrison: "STATIC, and the monotonous BEATING OF RHYTHM to a ROCK AND ROLL SONG, but by savages who have learned the words by rote. They sing 'Light My Fire' in the background." The original plot had Willard and Lance, the only surviving members of the PBR crew, join Kurtz's band—in some ways, literally, a band—to battle with the NVA. Kurtz asks Willard, "How do you like the Doors': 'C'mon Baby Light My Fire'?" and Willard shrugs and replies he likes the song. Kurtz gushes, "I love it."[21] (How unimaginable is this exchange between Martin Sheen and Marlon Brando in their roles as Willard and Kurtz?) Kurtz, like Kilgore, fights to amplified music, with "enormous loudspeakers protected behind spirals of razor-sharp concertina wire. LIGHT MY FIRE is blasted out to the enemy, poised to attack." Milius even included a technical exchange about making sure the "battle song" wouldn't be silenced. Captain Colby assures Kurtz he has rigged "dual tapes underground. If one is hit the other will continue to play."[22] In Coppola's 1975 draft, Kurtz's men, high on acid, speed, cocaine, and pot, listen "like those people you see listening to radios in their cars."[23] Dismissive antipathy toward rock fans sneaks into the scripts with this comment.

Apocalypse Now lacks a writer or director of the Vietnam generation who is trying to recover an experience that, however traumatic, was also part of the wonder of youth. The bitterness of Coppola's film, its refusal of sentiment and nostalgic music, marks it as an outlier in the normally warmly human genre of the PCF. Indeed, Kurtz's men in these script drafts take positively sensual pleasure in the spectacle of destruction. Willard, "exhilarated," moves to the music and says "napalm." The description of the mix emphasizes the reactions of the Americans: "The SCREAMS of maimed and dismembered men almost penetrates the INCREDIBLY LOUD MUSIC and we HEAR Kurtz's men LAUGHING and SCREAMING in delight," and through it all, "Blared out at tremendous volume over and above the DIN OF BATTLE is LIGHT MY FIRE," which spurs the Americans and the Montagnards on to "utter and most horrible savagery." The battle is the classic enemy-in-the-wire scenario—Kurtz's men cannot prevail against the NVA—ended by the Americans calling in an air strike on their own position, climaxing with "a spectacle of MUSIC and light and fire and overwhelming color," a "spectacle of total psychedelic war" accompanied throughout by diegetic psychedelic "HARD ROCK MUSIC."

Coppola and Milius's fundamentally musical vision of the Vietnam War was hardly realized in the finished film. The script drafts suggest a rather different film, where rock music is indicted as the sonic equivalent of America's suicidal embrace of violence and death in Vietnam.

SILENCED SWING IN THE WORLD WAR II CYCLE

A World War II–era cartoon in *Down Beat* magazine has an American tank crew rolling through the African desert debating the relative merits of clarinetists and bandleaders Benny Goodman and Woody Herman. Such conversations about popular culture are entirely absent in the World War II cycle of PCFs. Indeed, there's almost no swing music in these films at all. The men in *Saving Private Ryan* gather around a phonograph salvaged from the ruins but have only Édith Piaf records to listen to: here, popular music does not stand in for home. There are no traces of American popular culture in *The Thin Red Line* (1998)—not even a harmonica. Early in the first episode of *Band of Brothers* (2001), the swing tune "American Patrol" is heard on a radio in a pub, only to be switched off when an officer enters to address the company.[24] Similarly there's but one swing record heard in *Windtalkers* (2002), playing during a barracks poker game quite similar to the "Okie from Muskogee" scene in *Platoon*. (The film's only sustained musical thread—touching in context and a welcome relief from the frequently graphic violence of the combat scenes—involves a sequence of scenes culminating in a cross-cultural duet for Navajo flute and harmonica.) Popular music heard on radios in a few scenes in *The Pacific* (2010) far back in the diegetic mix never succeeds in penetrating the affective lives of the characters. In *Red Tails* (2012), music piped over loudspeakers at the base in Italy where the Tuskegee airmen are stationed—ironically it's presented as Axis radio—subtly shifts from sweet (whiter) swing to hotter (blacker) jazz as the film progresses, suggesting for the astute moviegoer the growing comfort of the black fliers. The men themselves, however, respond not at all to this music. In short, the aesthetics of swing as popular music, dance, and culture defining the World War II generation has not been exploited by PCF filmmakers—who are, themselves, mostly of the post–rock and roll generation. Men like Steven Spielberg, Spike Lee, and John Woo may simply not understand swing or hear it as an effective affective resource for their audiences. Or perhaps it's their parents' music.

The lack of popular music and culture also impacts how "cool" the soldiers of World War II appear to contemporary viewers. Composer Terence Blanchard's score for *Red Tails*, benefiting from the hip quotient always attending pilots, inserts an electric bass–driven, slightly funky beat for the cue "Takeoff." Later, on their return from a successful engagement with the enemy, Blanchard celebrates some fancy flying by the hotshot Joe "Lightning" Little with a jaunty march straight out of pre-Vietnam war film scores. Swing, apparently, wouldn't do in either case.

The absence of swing in these films suggests that the PCF cycle begun in the 1990s was not nostalgic but rather mythic, re-creating the 1940s without recourse to the popular music of the time, a strategy that lends a timeless, rather than period, quality to these often self-consciously authentic films. But as a result, these films denied their older audiences the chance to swing out one last time to the music they loved.

The only director of a World War II PCF to show American GIs enjoying popular music of the 1940s is Clint Eastwood. He's also the oldest director in this group by a generation—born in 1930, old enough to remember and claim the musical culture of the war years. Eastwood's knowledge of jazz surely plays a role here as well. *Flags of Our Fathers* (2006) uniquely incorporates a song scene into a World War II PCF. The scene also marks a structural transition in the film, coming just before the Iwo Jima landings, which are depicted in visually spectacular fashion.

Popular music is introduced by way of the shipboard intercom operator tuning in to a Japanese propaganda broadcast hosted by Tokyo Rose and piping the music throughout the ship. The central characters of the story are shown on deck enjoying the sound of a small-ensemble blues record. One Marine even plays along on his guitar. Paul Haggis's script calls for "big band music" here.[25] This more intimate record keeps the scale of the scene and of the music small, the circle of listening Marines as stand-ins for the music makers.

A 1944 recording of Dinah Shore singing "I'll Walk Alone" follows, introduced by Tokyo Rose as "this sweet music is to make you think of your girls back home who aren't missing you." The record hails from a strange interlude in popular music history. From mid-1942 to late 1944, the musicians' union refused to record, hoping for a cut of royalties on disc sales. Singers were not in the union, so they could record, but not with instrumental backing. And so the disc heard in *Flags of Our Fathers* features Shore's solo voice against a background of ooh-ing and ah-ing choral voices. Shore's record, faded in about halfway through its length, provides an oasis of calm during which the men, mentally preparing to fight the next day, reflect silently. The camera moves lyrically across their motionless figures. On a cut back to the dark of the deck, the Marine with a guitar strums a few chords with the record, the clear sound of his instrument adding a further diegetic layer to the slightly scratchy sound of Shore's recording. It's a beautiful sonic touch facilitated by the lack of instruments on the record, making Shore's vocal presence into something almost palpably real. The words she sings are about faithfulness, the opposite of the portrait Tokyo Rose offered of the women back home. Shore's a stand-in for those American girls, indeed for home itself. This sweet music of the 1940s—a brand of popular music conventionally favored by female listeners—here gives the men the chance to sit quietly with their thoughts, to reflect without talking on a topic otherwise untouched in the film. The song scene ends with a quiet click: just after the final cutoff of the record, a soldier reading a letter by the light of his Zippo clicks the lid shut. Like the cough at the close of "White Rabbit" in *Platoon*, Eastwood uses this bit of sonic punctuation to kick off the grand spectacle of the landing on Iwo Jima.

Most audiences for *Flags of Our Fathers* likely register Shore's record as authentically old and attend mostly to the lyrics, which speak appropriately to the situation. Those who lived through the war probably recognize both song and singer. Either way, the musical moment works. Only Eastwood has exploited the resource of wartime popular music in a World War II PCF, and the choice of song and record was likely his. Haggis's script does not mention "I'll Walk Alone," and Eastwood sings a lonely phrase of the song at the very opening of the film, his unaccompanied voice in the dark the first ghost of the past raised to life by the film.

UNREPRESENTED MUSICAL PRACTICES IN POST-9/11 PCFS

Soldiers use music differently in the twenty-first century, in part because the digital revolution makes bringing music to the battlefield so much easier. As J. Martin Daughtry notes, "The iPod's timeline coincides with that of the Iraq War to an uncanny degree," and "years of active use within the military rendered the iPod an icon, index, and, to a significant extent, militarized instrument of the war."[26] At the very least, iPod warriors don't face the problems of soldiers like Motown in *Hamburger Hill*, whose cassette player runs out of batteries and sparks a fight between black and white.

And yet popular music is generally absent from the GWOT cycle. One explanation may be filmmakers' discomfort with the ways the US military has used popular music as a tool of torture—or "enhanced interrogation."[27] A scant five seconds of the track "Pavlov's Dogs" by the group Rorschach is heard during the torture sequences early in *Zero Dark Thirty* (2012). The use of popular music to break down detainees, a well-documented practice, is more referenced than re-created in any affective way for the viewer.

The widely attested use of music played at high volume over loudspeakers to prepare soldiers to fight or to assault the enemy has been completely excluded from GWOT PCFs—surprisingly so, given soldiers' witness. Colby Buzzell lovingly details the improvements his crew made to their equipment, especially a "loud speaker on the outside of the vehicles. Nothing motivates troops more before a mission than good motivational music. Remember in the movie *Apocalypse Now*, when they had the speaker hooked up to the Air Cav helicopters? Well, we need to do the same goddamn thing with the Strykers." He describes the effect of music on a joint mission with Iraqi Civil Police. "Anyways, while we were driving out the main gate to the FOB to do this joint mission we had the loudspeakers blasting 'Ride of the Valkyries,' [the] theme song from *The Good, the Bad, and the Ugly*, 'The Star-Spangled Banner,' and the *Rocky* theme song. It motivated the hell out of us all. In fact, it even motivated the Iraqi Police, I remember looking over at them and they were all getting into it big time."[28] The battle of Fallujah included sonic wars between American and Iraqi loudspeakers. Bing West reports how psyops crews had loudspeakers instead of machine guns on the roofs of their Humvees. West heard such crews "playing at full blast [Eminem's] 'Let the Bodies Hit the Floor'"—West misidentifies the song's artist; it's actually by the band Drowning Pool—and "blasting Jimi Hendrix at 110 decibels."[29] Other songs on the psyops playlist were Guns 'n' Roses' "Welcome to the Jungle" and, a frequently mentioned record, AC/DC's "Hell's Bells." West completes the sound image by including the Iraqi side's response: "Not to be outdone, the mullahs responded with loudspeakers hooked to generators, trying to drown out Eminem with prayers, chants of *Allahu akbar*, and Arabic music. Every night discordant sound washed over the lines." West reports that psyops also drew on action-horror film sound effects: "The top chiller was the deep, sinister laugh of the monster in the movie *Predator*, played in low bass at one hundred decibels, echoing off the pavement." An American crew farther up the street—unnerved themselves by the laugh in the night—asked them to stop. On the Marine Corps' birthday, psyops played "The Marines' Hymn." West describes the soundscape of the city in positively cinematic terms, not far from Coppola and Milius's evocations of Vietnam only with twenty-first-century and Middle Eastern sources: "Fallujah after dark was a cacophony of sounds: dogs yapping and howling, explosions near and far, bursts of small-arms fire, the annoying whine of the Predator UAV and the rumble of Slayer [the heavy metal band] overhead, and high-pitched calls from the minarets to evening prayers."[30] No GWOT film has attempted to re-create this potent sonic stew. (Nor, for perhaps obvious reasons, has the numbing roar of generators in the "industrialized soundscape" of Baghdad described by Daughtry been evoked in any PCF soundtrack.)[31]

GWOT soldiers' personal use of music is well documented. Buzzell's 2005 *My War: Killing Time in Iraq* is an especially rich source. For instance, Buzzell introduces his buddies by detailing their musical tastes: "a brother from Baltimore" who is a "hip-hop/rap enthusiast"; "an old-school headbanger from back in the day who is way into metal and thrash" and married with two kids; a medic "into punk and alternative music" who "always carries a *Rolling Stone* or *Spin* magazine." Buzzell himself prepared for war by buying an iPod: "I downloaded a 'We're All Gonna Die' mix on my iPod for the party. It was just basically a bunch of songs about war and death and getting killed. (Mostly by Slayer.)"[32] Buzzell also lists his "My War Soundtrack," which includes the Metallica track "Kill 'Em All" (a title that resonates with the discussion of that phrase in the previous chapter) and, perhaps surprisingly, Louis Armstrong's "What a Wonderful World" (clearly he knows *Good Morning, Vietnam*). And Buzzell is no exception. As Lisa Gilman demonstrates in *My War, My Music: The Listening Habits of U.S. Troops in Iraq and Afghanistan* (2016), "Musical listening was one of the few activities over which U.S. troops had control during their deployments and that they could do both privately and collectively. For many service men and women, listening to music provided one of the most salient mechanisms for thinking, feeling, escaping, communicating, connecting, passing the time, bonding, hiding, and grieving."[33]

There's ample evidence for the varied ways in which soldiers—and the military officially and unofficially—use popular music. Yet seldom have these contemporary practices been pictured more than very briefly in GWOT PCFs. Indeed, a majority of GWOT PCFs include none or very few licensed American popular music tracks among their end-titles clearances: *United 93* (2006, none), *In the Valley of Elah* (2007, seven), *The Kingdom* (2007, two), *Redacted* (2007, three), *Green Zone* (2010, three), *Act of Valor* (2012, none), *Zero Dark Thirty* (six), *Lone Survivor* (2013, one), *American Sniper* (2014, one). None of these films include main characters intentionally listening to or enjoying popular music. All—with the exception of *Lone Survivor* and its new-guy ceremony—use their minimal compiled cues as diegetic background noise, often as part of a public space such as a bar, frequently moved through briefly. If the World War II cycle's avoidance of swing renders the citizen soldiers of that era as figures in a quasi-mythic past, the absence of contemporary popular music in the GWOT cycle sets elite professional warriors of the all-volunteer force apart from other Americans, outside the stream of popular culture. These men (and two women—played by Jennifer Garner in *The Kingdom* and Jessica Chastain in *Zero Dark Thirty*) have no musical tastes to speak of, a lack that limits them to their professional functions and, to some degree, narrows their humanity.

Only *Generation Kill* includes GWOT soldiers as embedded in popular music and as themselves musical. These "young Americans unplugged" do not have access to iPods or other audio equipment, and so they sing instead. Very short, diegetic musical snippets—about twenty-five moments of men bursting into song—prove series highlights: a montage of all the singing in *Generation Kill* is usually up on YouTube. The Recon Marines' use of music is specific to their situation and frequently mocking, almost all of it initiated by the driver Ray Person. The group's repertoire represents an eclectic mix of styles: from pop (Avril Lavigne) to metal (Drowning Pool) to hip-hop (Nelly, Tupac, Dr. Dre, N.W.A.) to rock (Styx from the 1970s; Wheatus from the 1990s) to the in context rather random 1970s song "Loving You." One Vietnam record—"I Feel Like I'm Fixin' to Die" by Country Joe and the Fish—shows up as well (also featured in *Hamburger Hill*). Teasing use of country music runs across the series: Person generally sings country to irritate Sergeant Brad "Iceman" Colbert, who dismisses Toby Keith's post-9/11 country anthem "Courtesy of the Red, White, and Blue" as "Special Olympics Gay." The book quotes the real Colbert calling country "the Special Olympics of music."[34] There were, apparently, no genuine country music listeners in the group profiled by the journalist Evan Wright. Still, Person and Colbert's Humvee crew revel in sing-alongs of "On the Road Again" and "King of the Road."

In his book, Wright describes singing as a "combat-stress reaction": "A lot of Marines, when waiting for minutes or hours in a position where they expect an ambush or other trouble, will get a song stuck in their heads. Often they'll sing it or chant the words almost as if they are saying Hail Marys."[35] The series doesn't use music in this way but instead places bits and pieces of songs into the flow of talk, which often turns toward pop-culture matters. Left and right are represented, with, for example, critical discussions of Disney's *Pocahontas* (1995) that are at once playful, serious, and articulate. Early in the series, one black soldier says, "I don't do racial stuff." Stafford, a white Marine, proves the most invested in musical blackness—specifically rap—and racial critique. Another Marine examines the ethos of the group and asks, "Ever realize how homoerotic this stuff is?" Drawing closely on a journalistic source, *Generation Kill* uses soldiers' song as a key element in its presentation of a racially integrated, post-hip-hop military. Musical taste isn't racially contentious but instead a matter of class and participation. The unreleased film *American Soldiers* goes further, including a lyrical interlude inspired by one soldier's admission that he can't dance. An interracial group forms a circle to coach him to an original hip-hop-style beat provided by a member of the squad known to beatbox. The men enjoy the moment like kids. In her profile of an especially hard-hit unit of American soldiers in Iraq, the journalist Kelly Kennedy describes soldiers practicing salsa dancing with each other and learning new moves while deployed. One soldier—described as "white as they come" who "looked as if he should be at the local cowboy bar"—especially loved to salsa.[36] War correspondent Richard Engel recounts a squad in Iraq who burst into dance every time the Black Eyed Peas' "My Humps" played on their common playlist. Soon, other squads were dancing to the tune as well in a kind of compulsive ritual.[37] The men who fought in the Iraq War hail from a fragmented and mashed-up popular culture scene. The monolithic style differences between black and white music in the 1960s are gone, especially in a movie context where we don't hear recordings but rather fragments sung by the soldiers themselves for one another. Only a viewer steeped in popular music of 2003 is likely to get all the references in *Generation Kill.* So while the series uses a lot of popular music for a GWOT film, it does not have the kind of nostalgic punch that 1960s music has in the Vietnam cycle.

It's hard to sing metal—there isn't much of it in *Generation Kill*, just a fragment of "Let the Bodies Hit the Floor" sung as "Let the *hajiis* hit the floor"—and yet metal seems to be a battle music of choice for contemporary soldiers. Jonathan Pieslak's study of the personal musical habits of soldiers in Iraq documents how individuals and squads listen to metal to pump themselves up before going outside the wire.[38] Pieslak even speculates that "the correlation between this music and combat, which is felt by many soldiers, derives from how the music re-creates the sense of gunfire and shooting an automatic weapon. . . . The element of inspiration may be created through the sense of power a soldier feels when firing an automatic weapon."[39] Daughtry writes, "The sonic onslaught of metal, arguably the most popular genre among US troops, felt liberatory and transgressive, but also safe: it was predictable, rhythmic, 'noise'-infused music, performed aggressively, composed with catharsis in mind—and remaining at all times under the listener's control."[40] Soldiers' use of metal as preparation for facing the enemy, again seemingly an invitation to cinematic re-creation, has simply not been put into films. Metal is heard briefly twice in *The Hurt*

Locker (2008), but only during downtime inside the wire, when Sergeant James listens to tracks by the metal band Ministry. In both cases, listening to metal is presented, somewhat counterintuitively, as a means to relax. (A third Ministry track is incorporated into the score's final narrative cue—see chapter 11.)

What is lost in the general absence of twenty-first-century soldiers' use of popular music in the GWOT PCFs? Obviously the US military's systematic use of popular music as a weapon to trouble the enemy and to break down detainees is elided. More to the point of the subgenre's historical function as a bridge between the American audience and the American fighting man, the identity of these soldiers is denied a grounding in popular culture, making them less legible as part of a shared and varied space, segregating military from civilian—a widely remarked broad cultural result of the all-volunteer force—and minimizing regional differences among the men. The post-9/11 force in the PCF is without popular music characterization: a lack that allows these fighting men to fade into a kind of cultural sameness, cutting off potential points of shared identification between specific characters and individuals in the audience, effectively abandoning a tool of shared experience richly exploited in the Vietnam cycle.

THE LOSS OF PATRIOTIC SONG

In an article on *The Deer Hunter* (1978) in *Time* magazine from the height of the media discussion of the film, a professor of religious studies opined, "Viet Nam means that patriotism can never again be understood in the simple way it was before."[41] The use of Irving Berlin's "God Bless America" in *The Deer Hunter* suggests the same might be said for patriotic song.

The most famous moment of singing in any PCF happens at the very end of *The Deer Hunter*. Having returned from Nick's funeral, the surviving friends—male and female, including two veterans: the wheelchair-bound Steven, Michael in his dress uniform—gather at John's bar. While making eggs for the group, John begins to hum "God Bless America" as a hedge against tears. Linda, Nick's fiancée, picks up the strain and the group sings it together. When the tune ends, they look at each other in silence, then raise a toast "to Nick." The film's gentle guitar and strings theme music comes in for a final visual roll call, including a happy, smiling Nick.

Perhaps because it was the last moment of a long, exhausting film, almost every review of *The Deer Hunter* had something to say about the singing of "God Bless America." It is the most commented upon musical moment in PCF history. The following digest of quotes from reviewers reveals how familiar songs sung in specific movie contexts can elicit complex, contradictory, productively ambiguous responses:

- "The survivors end up singing 'God Bless America,' and they mean it. But it makes all of us reflect upon the price we paid for a war that few of us wanted. . . . Unless I am very much mistaken, it is the affirmation of an ultimate belief in this country. These young men have ventured beyond our borders and witnessed at first hand the savagery and corruption that rule there. Their 'God Bless America' is fervent and heartfelt. If we find it ironic, we'd better, goddamned well be able to spell out why."[42]

- "Is 'God Bless America' at the close an affirmation after all? Yes, the studio synopsis suggests helpfully. I wonder, says the viewer, sensing—simply in the context of the movie itself—a bitter irony. Ambiguous, says the verdict: inconclusive either way. Perhaps the movie is intended to be, as it sees the war to be, a cleansing, a purging of old attitudes, achieved at hideous cost in death, disfigurement, disillusion, but achieved nonetheless and providing a basis for a calm and rational future. Perhaps."[43]

- ["God Bless America"] "is a dangerous and courageous move by Cimino, one that will embarrass, even anger, many sophisticated minds. The scene is not perfectly done, perhaps cannot be, but it is a moving statement about the spiritual danger of rejecting imperfect community for what may be an even more imperfect righteousness."[44]

- "It is a bold dramatic stroke, and in the hands of a less gifted director or cast, it might have provoked titters from the audience—or even guffaws. But Cimino was wise to take the risk, for he succeeds in tapping a profound, if profoundly buried, emotion: a basic yearning to retrieve some semblance of national pride, or at least some sense of fellowship and community, in the wake of the harrowing debacle of Vietnam. The scene is done with sympathy and tact: it is no exhortation to mindless patriotism. . . . The need for restoration is also potent, however, and it is something to which none of us is finally immune."[45]

- "It is a scene that, in lesser hands, might have been played for a cheap and easy irony. I don't think that any irony was intended . . . or if it was, it's the kind of irony that forces the viewer to examine the reasons behind his own impulse to laugh."[46]

- "[*The Deer Hunter* ends with the surviving characters] singing the familiar words in a setting that makes them new. The film ends on this note of self-forgiveness. More than a display of wrongheaded patriotism, the singing represents an acceptance of terrible accusations coupled with a resolve to carry on, to choose life over death. We plead guilty, but affirm our basic moral strength to live."[47]

- "I submit that, if we are going to be moved to thought and action by *The Deer Hunter*, it ought to be by the implications of its true subject: the limitations for our society of the traditions of male mystique, the hobbling by sentimentality of a community that, after all the horror, still wants the beeriness of 'God Bless America' instead of a moral rigor and growth that might help this country."[48]

- "Such is Cimino's fresh perspective that *The Deer Hunter* should be an equally disorienting experience for hawks and doves. This is the first movie about Viet Nam to free itself from all political cant. . . . But the film's ending, in which the major characters spontaneously sing 'God Bless America' at a funeral breakfast, may give audiences some pause. The moment is powerful, all right, but does one laugh or cry? It is hard to do either. Like the Viet Nam War itself, *The Deer Hunter* unleashes a multitude of passions but refuses to provide the catharsis that redeems the pain."[49]

- "As I watched the 'God Bless America' conclusion, feeling slightly sickened by Cimino's avoidance of a moral statement, I remembered a high school friend who left home the same time I did. I went to college. He went to Vietnam. We were friends, but we had argued—I enthusiastically, he reluctantly—about the war. I came home at Christmas in a jet. He came home in a shoe box. Hank was serious in his support of what we called the U.S. involvement. He has been dead for 10 years. Now, a movie is weeping for him and for the thousands like him. It weeps in a way he, and they, would understand. One does not have to agree with *The Deer Hunter* to sympathize. One does not have to like it to recognize its value."[50]

These reactions to *The Deer Hunter*'s closing song suggest the power of patriotic song to trigger deeply ambiguous meanings. The significance of the moment—combining familiar content with surprising, difficult-to-read diegetic performances by the characters—defied easy response but demanded explanation. Perhaps Cimino gets close to the intended effect in his description of the scene in the script: "They all seem caught in the

intensity of the moment, but whether in joy, relief, or for some other reason, we cannot tell. All of them there and the singing are like one thing, a
thing inevitable, older than the memory of man."[51] The fraught nature of
the Vietnam War as remembered scarcely four years after the fall of Saigon
resonates powerfully in these quotes, reminding us that the founding PCF
cycle had profound cultural work to do. The Vietnam cycle mattered in
ways that it is difficult to remember or imagine now.

And yet, interpolated patriotic song is absent from almost every post–
Deer Hunter PCF. "Grand Old Flag" in *Born on the Fourth of July* scans
more as a reference to Kovic and an alternative to "Yankee Doodle Dandy,"
from which the film's title is drawn, than as a freestanding patriotic salute.
Only *Red Tails*, a visual effects–oriented film with specific racial work to do,
makes direct use of a standard patriotic anthem: "America the Beautiful"
sounding in a jazz arrangement for studio orchestra over its end titles. This
remarkable absence of patriotic song in war films for the last forty years—
even those offering unambiguous support for the military—marks the tremendous legacy of Vietnam, enduringly able to trouble easy markers of
American pride, at least in the PCF as an arena for patriotic reflection.
Other musics have stepped into the breach (see part 4). The result is an
opening outward of the cinema as a space where nuanced versions of patriotism and a movie music tuned to mourning the costs of war might be
expressed and experienced.

5. Disembodied Voices

A soldier's voice raised on the battlefield reveals the presence—and suggests the location—of the soldier's body.

Sometimes such cries are intentional, inviting attack. The only representation of the enemy in *Apocalypse Now* (1979) is a sonic taunt. (Coppola's film follows George Ball's statement in a famous [and ignored] memo to Lyndon B. Johnson: "We cannot win, Mr. President. . . . The enemy cannot even be seen in Vietnam.")[1] At the Do Lung Bridge, an unseen Vietnamese soldier taunts the American front line, repeatedly shouting, "Fuck you GI." Roach, a fearsome black American soldier, shoulders his grenade launcher—painted in tiger-stripe camouflage—takes aim, and fires into the blackness. The enemy voice falls silent. The film's noisy soundtrack, especially dense and varied in this episode, goes eerily quiet around this exchange. Sound designer and rerecording mixer Walter Murch noted of this soundscape, "At least for this brief period in the film, you are hearing the world the way Roach hears it, which is focusing in with a sublime subjectivity on just what he needs to hear."[2]

Other times, cries heard and followed on the battlefield draw soldiers toward a revelation of the effects of their actions. In *Born on the Fourth of July* (1989), just after Kovic and his men fire on a Vietnamese village, they hear "voices screaming in the distance."[3] The screams draw them to a hut, where they find they have mortally wounded several women and children. Now the American soldiers start screaming and crying, adding their lamentations to those of the dying.

The cries of the unseen might also belong to comrades who fell in no-man's-land. Among the most moving sections of the 1949 combat film *Sands of Iwo Jima* is a stretch where John Wayne and John Agar listen to the screams of a wounded American who cannot be rescued. The hardness required to resist responding serves as the lesson Agar needs to learn from Wayne. A similar, though differently nuanced, sequence in *The Thin Red Line* (1998) has a medic rush out to help a gravely wounded, loudly screaming soldier only to fall himself, instantly shot dead. Captain Staros worries that the still-screaming man's cries will take all the fight out of his remaining men. Sergeant Welsh dashes to the wounded man through fierce fire—noisy effects and noisy scoring attend his run. Welsh can do nothing but give the wounded soldier sufficient morphine to end his own life and, thereby, quiet his own cries.

The discipline *not* to cry for help—to resist revealing one's presence and location—must be taught to soldiers new to combat. Sergeant Elias, early in *Platoon* (1986), says to several new guys before heading into the jungle on a night patrol, "If you get lost, don't cry out. We'll find you." How?—their faces collectively ask.

For medics, of course, cries for help—voiced as "medic" (by soldiers in the Army) and "corpsman" (by Marines served by Navy medics)—drive their combat action. The medic Eugene Roe in episode six of *Band of Brothers* (2001) races from position to position on the line in the Ardennes Forest, answering to the call "medic" voiced by frozen troops during the Battle of the Bulge. The artifice of film sound—a synthetic creation, after all, amenable to the representation of soldiers' inner lives—allows such cries to resound across time. *Flags of Our Fathers* (2006) includes echoing calls of "corpsman" as bridges between the film's several time frames and opens with the young Doc Bradley running through an empty battlefield—a moonscape evoking the subjective experience of combat—looking for his lost battle buddy, crying out, "Iggy." The voice of the young Bradley awakens the old Bradley, an aged veteran in his bed, still hearing his own fruitless cries, his younger self still calling out to his older self, hoping to find a young soldier who was, when the original calls were made, already dead.

This chapter considers the full range of disembodied voices in the PCF, from diegetic voices heard by way of technology to standard movie voiceovers, adjusted to the unique needs of the subgenre, to letters spoken as voice-overs by the living and the dead.

ON THE NET

Modern combat is not defined by a natural acoustic. Electronic communication technologies—radios, phones, integrated CNR (combat-net radio) systems—have slowly transformed the battlefield into a realm of technologically mediated disembodied voices. These communication technologies provide a structuring sonic armature for many PCF narratives, opening a contiguous sonic space where combat coordinated across great distances can unfold in an orderly, or at least comprehensible, fashion. Film proves an ideal (and idealized) medium to render visible the complexities of the communications net, lending a kind of intelligibility to the battlefield (that likely exceeds the realities of the contemporary American military in some GWOT films).[4] This aspect of war film narration often puts characters in the position of listeners while the film viewer usually enjoys privileged visual access to all parties on the net. As a result, the image of a soldier listening to a disembodied voice is common across the PCF. At such moments, the film viewer is doubly at work interpreting both what the listening soldiers hear and the reactions on their faces to what they are hearing.

The RTO (radiotelephone officer) is, by default, a central character in war films set in Vietnam. He carries the unit's communications capability on his back, usually sticking close to the nearest officer and connecting the men on the ground to the full resources of the military machine supporting them. As veteran author William Pelfrey puts it in his 1987 novelization of *Hamburger Hill*, "The calm radio-telephone operators . . . were the umbilical cord between the troops in the bush, the helicopters that might or might not take them out, and the officers monitoring their casualties and kills back at battalion, brigade, and division command posts."[5] In most World War II films, communication is by phones connected by wires to the command post, the lines always susceptible to breaks. As with RTO units in Vietnam, World War II phones limit the number of soldiers who can speak and hear on the line. Grunts in both conflicts sometimes overhear one side of conversations that might determine their fate. The heated exchange between Staros and Tall in *The Thin Red Line*—the former formally defies the latter's order to send his men into direct fire—is overheard by Staros's men. They listen, hoping Staros stands firm. He does. They thank him in a subsequent scene.

As technology improved in the post-Vietnam era, the occupational specialty of the RTO was eliminated. Thus the RTO does not appear in PCFs set in the Middle East. Indeed, from *Courage Under Fire* onward most every soldier is on the improved and pervasive combat-net radio, vastly increasing the cast of characters for what becomes a more complicated combat radio drama. Putting everyone in the same sonic space emerges as a key narrative device in most all action-oriented Iraq and Afghanistan films to follow. These are films about soldiers connected by headsets, which become essential lifelines. In *The Hurt Locker* (2008), after James removes his "cans" during an especially tough bomb-defusing challenge, Sanborn socks him full in the jaw: "Never turn your headset off again," he says, angrily. Curiously, contemporary SEAL equipment, such as Peltor-brand headsets that enhance hearing and moderate the impact of loud noises, has not been represented.[6] The SEAL as a kind of cyborg warrior, frequently celebrated in military literature, has not been a theme in the PCF, which instead insists on the underlying humanity of these elite fighters and the importance of their bodies. The technical niceties of the contemporary CNR are less important than simply being on or off the net.

PCFs represent CNR selectively, as serves the story being told. In *Black Hawk Down* (2001) the viewer sees everyone on the net: from the bottom to the top of the command chain, from Garrison at the TOC to observers in helos circling above the city (who never break a sweat) to individual soldiers on the streets of Mogadishu. The first combat death in the film is reported on the net, setting up a series of reaction shots of men at all corners of the battle. In *Generation Kill* (2008), visual representation of the net stops at General Mattis: his commanding officers remain unseen even when he talks to them. The viewer is effectively embedded with the Recon Marines throughout and never gets a broader sense of the invasion, a crucial limiting of the series' scope that restricts any larger narrative of American victory. The series also develops a sense of the net as a radio show being listened in on by the men themselves. We hear the comms—official chatter—and the reactions of men listening to what's said on the comms—the scuttlebutt of side comments to buddies nearby.

There are economic advantages to storytelling via the net. At some point, most American combat stories appeal to the skies—to bombers waiting to drop their loads or to planes facilitating encrypted wireless communications. The easiest way to represent such planes narratively is to show the image of an aircraft—an effects or stock footage shot—accompanied by the voices of these planes' unseen pilots on the net. This happens in *Platoon* and *Lone Survivor* (2013).

On a more substantive thematic level, acousmatic battles—battles heard but not seen—are common, especially in the Vietnam cycle. At such moments we listen to the sounds of the dying over the radio while watching our characters listen as well. Often these moments come before the listening soldiers face the enemy themselves, creating dread of the anticipated combat. J. Martin Daughtry terms such experience in actual combat "*belliphonic audionarrative*: the story of an unseen battle unfolding before one's ears."[7] The squad in *84 Charlie MoPic* (1989) listens to a battle on their radio and briefly debates whether to go to the struggling unit's aid. (They don't.) On a training run at their stateside base before deployment, officers in *We Were*

Soldiers (2002) listen in on a battle happening in Vietnam by way of radio signals bounced off the atmosphere. *Platoon* opens its final combat sequence with a short battle heard by radio. A pacing Captain Harris—played by retired Marine officer Dale Dye, his RTO following as if on a leash—tries to help a unit under fire where all the officers have been killed.[8] As described by Stone, "A young inexperienced VOICE screams back into the radio amid intense background FIRING filtered by radio and sounding disembodied."[9] The drama of the moment lies in Harris's engagement with the soldier on the radio. He calmly calls the young man "son" and tries to talk him through but doesn't get very far. Soon there's no response. The connection reopens to extreme noise, then the sound of spoken Vietnamese, then complete silence. Harris responds with a single word—"shit" (a line not found in any script, perhaps ad-libbed by Dye)—which, like a single-syllable beat of punctuation, kicks off the next phase of the combat sequence (analyzed further in the next chapter).

Radio chatter need not carry plot information; it can work, as Murch suggests of sound effects, like music. Episodes of easygoing Code Talker chatter in Navajo (with subtitles in English) punctuate the structure of *Windtalkers* (2002), adding to the film's characterization of the Navajo Marines as gentle spirits unlike the generally angry or warlike white Marines. *Generation Kill* uses only radio traffic for its opening and closing titles and throughout the series as a kind of scoring over transitional images of the Recon Marine convoy on the move. Some of these transmissions carry story content, some do not. All are measured, restrained, and professional. Radio chatter is a realm of specially coded talk that can fall on the uninitiated ear as so much static-y verbal music. But code language in any PCF better be correct on some level to satisfy military audiences. And once you know any call to the "six" is a call to a commanding officer, the meaning of many radio transmissions heard in PCFs opens up considerably.

Creating the sound of the net is, of course, of a piece with making the soundtrack as a whole. Voices heard on the net have to be recorded and mixed just as with any actor's performance. To arrive at authentic chatter for the helicopter attack scene in *Apocalypse Now*, sound re-recordist Richard Beggs invited four veteran helicopter gunship pilots onto a soundstage. He also invited their wives. Using army surplus headsets with microphones, Beggs ran the film without sound: "What I did do was to pump into the room via these giant JBL studio monitors—at ear-splitting levels—the interior helicopter sounds of the Huey. So the air was almost rarefied by this loudness, and they're watching this image, and after about a half an hour or so, they're screaming at each other, and reliving it."[10] Beggs reports the line "get that dink bitch" was improvised by one of the pilots in the presence of his wife, who broke down in tears.

Phones and cell phones extend the CNR all the way home. *Black Hawk Down* cuts away from Somalia only once: just before heading out, a pilot who will be dead by the end of the day calls home. He leaves a message on the answering machine. His wife walks into their sunlit kitchen just as he clicks off. Both are left listening to recorded voices (see below for the trope of tape delay). In subsequent PCFs, the possibility of hearing disembodied voices from home is similarly framed as a missed connection—even when the connection is made. Swofford calls his girlfriend Kristina in *Jarhead* (2005) from a bank of phones in Iraq. Their short, difficult call leads to a quasi–dream sequence scored with the Nirvana track "Something in the Way," a resonant lyric for the moment. Chased by the Iraqi army and trapped in a room of stolen Western goods in *Three Kings* (1999), Troy picks up a cell phone and calls home. His wife answers in the clutter of daily life with small children. This glimpse of America sets up later imagined scenes of Troy's house and family being destroyed by Iraqi bombs, visions triggered in Troy's head by his Iraqi interrogator. Finally, *In the Valley of Elah* (2007) begins with a soldier's call home from Iraq. Parsing the meaning of the call, getting to the heart of his son's distress, serves as the film's driving obsession. As Stacey Peebles notes, "These communication technologies that have the potential to bring soldiers together virtually with their loved ones are shown, in *Elah*, to be more revealing of personal and social fragmentation than they are of any kind of union."[11]

The SEAL Team cycle tightens the connection between the war zone and the home front even further. *Lone Survivor* opens with Matt "Axe" Axelson chatting online with his wife. They exchange everyday talk: "gotta pay the bills," he says of the mission where he will die. The image track never cuts to Axelson's wife at home on her computer, but in the background in his bunk are photos of the real Axelson's wife. The pictures reappear in the film's final montage. In effect, Axelson's wife plays herself by way of real photos and online chat—a disembodied voice rendered as text. *American Sniper* (2014) develops a further innovative thematic and structural use of cell and satellite phone calls between husband and wife. Kyle makes several calls home while on the job—literally while staked out as a sniper. The calls always intercut images between Kyle and Taya, concretizing both parties in their lived environments for the viewer, who occupies a position of considerable access neither character enjoys. These calls allow for a dramatic representation of the fear families have for soldiers' safety. In one scene, Kyle and Taya are on the phone when he comes under heavy fire. He drops the phone—without disconnecting—and Taya is left listening to the battle. Here, a wife holding down the home front is put in the position of the soldiers in *84 Charlie MoPic*, *Platoon*, and *We Were Soldiers*, listening to an acousmatic battle, helpless to do anything. *American Sniper* ups the tension by having a pregnant Taya emerging from a doctor's appointment at just this moment and adds to the immediacy of the call by setting both ends in broad daylight (Iraq and San Diego are ten hours apart, making simultaneous daylight in both places uncommon).

TAPE DELAY

The above examples involve disembodied voices heard by characters in a film and experienced in real time by way of communications technology, a historically changing element of the modern battlefield that shapes the sonic texture and narrational techniques of the subgenre. Another set of disembodied voices sound by way of tape delay, where an audio storage medium brings absent others into a film by way of their voices only. Here, and with even more intensity, characters listen while the film audience both listens along and watches characters listening.

Letters from home in the form of cassette tapes figure prominently in two PCFs. The first death on the PBR (patrol boat) in *Apocalypse Now* is attended sonically by the dead man's mother—uncomprehendingly present on tape. Clean, played by Laurence Fishburne, briefly listens to a tape from his mom just before the boat comes under heavy attack. He leaps to his weapon as gunfire drowns out the tape. Clean is shot dead by the unseen enemy. After the firing stops and the soundscape clears, Clean's mother's steady, affectionate voice reemerges on the still-playing tape, heard in counterpoint to the shocked face of Chief, the boat's pilot, who looks directly at the camera for an interminable stretch. Clean's mother—almost the only woman's voice in the entire film—says, with humor, "stay out of the way of the bullets and bring your hiney home all in one piece because we love you very much" over a cut to the image of Clean's bloody body. The tape clicks off as Chief embraces the dead soldier and begins to sob, saying, "So young." It's the only moment of emotion and care permitted the otherwise brusque Chief. While it's unlikely that Chief and the others on the PBR are attending to the tape, for the viewer Clean's mother's words are eloquent—and, importantly, not ironic. The film represents the pain she will feel when she learns of her son's death by indirection: we hear her love for her son as we look upon his mangled body; we are put in the uncomfortable position of knowing more than she does. (This economical insertion of a soldier's mother takes *Apocalypse Now* a step before an analogous moment in *Saving Private Ryan* [1998], when Mrs. Ryan is initially shown washing dishes, then involuntarily falls to the ground when she sees a Casualty Notification team—a uniformed soldier and a minister—approach the front door of her Iowa farmhouse.) Clean's mother is apprehended as a disembodied voice that is, at this moment, all hope and no knowledge. If there is a heartbreaking moment in the otherwise cold *Apocalypse Now*, this is it.

A similar tape from home in *Hamburger Hill* (1987) works more specifically on the men listening while also challenging viewers' personal history in relation to the Vietnam War. The FNG (fucking new guy) Joe Beletsky's girlfriend Claire sends him a tape, which he listens to on Motown's tape player where a group of others can hear it as well. Claire's voice is initially futzed, mediated by the player's crude speaker, sounding far away. She speaks of time spent with Beletsky's parents and suggests, "If you can, Joe, try to smile more in the pictures [you send home]. It would help your mother." Claire tells him his father is following the war closely, "Eats his dinner in front of the TV every night. I think he's looking for you." Then Claire moves to sexual matters: "I'll always be true to you. I'll wait for you," she says, adding, "If you can't wait, if you can't be true, I'll understand," reversing the persistent PCF theme of women back home taking up with "Jody." By this point, all the men are listening closely and the film has come to a point of tense stillness. Claire's voice slowly transitions out of the cassette player and into the clearer, more intimate, more fully present non-diegetic space in the mix where voice-overs are placed. She begins to speak into our ear, to take on a palpable presence. By the end of the tape, she is right there. Claire begins to inadvertently address all the men, saying with rising emotion, "I wish you weren't there. I wish none of you were there. But I'm proud of you, Joe.... I don't believe what they say about you.... Because I love all of you." Claire is, without a doubt, the perfect PCF girlfriend. The impact of her sonic entrance into the film, so effectively mixed to make her presence increasingly felt, works to make any viewer who themselves once believed "what they"—by implication those protesting against the war—"say" feel uncomfortable. Viewers who did not live through the war years are unlikely to register Claire's soft-spoken accusation.

Script and film differ as to the sound of the tape and the presence of Claire. The script describes the moment this way: "The young girl's VOICE seems to penetrate every fighting position; the Troopers listen, not to what she says because that belongs to Beletsky. They listen to the young woman's voice that speaks without the stress and tension and profanity that has become their lives ... the voice gets weaker and weaker (the batteries are dying)."[12] Dying batteries would distort Claire's voice, slowing it down grotesquely. In the film, her voice remains clear, slowly taking over the soundtrack, pushing the film's normative noisily realistic diegetic mix into a slightly stylized, breathless realm. Then, abruptly, the machine goes dead. This is not how battery-run tape recorders work, but in the world of *Hamburger Hill* it's more important to preserve Claire's purity of sound and sympathetic voice than to reproduce the actual properties of a cassette player. The tape sequence crucially sets up a series of linked events. An argument over the recorder—Beletsky says Motown ran it down playing "his nigger music"—leads toward a kind of racial reconciliation, when, as described earlier, Doc greets Beletsky with a modified dap and the words "brother blood." A conventional "Dear John" letter—the bad-news kind—received by a heartbroken FNG and Frantz's confrontation with a reporter follow. Both episodes present the men's bitterness at their treatment by civilians, all set in relief by Claire's words of love to them all.

In a reverse example—tapes and video from the war zone sent home—*In the Valley of Elah* follows Mike's father's effort to recover and understand the video files and photos on his son's cell phone. He hires a tech guy to recover the files, and one by one receives short videos in his own email from Mike's phone. These unintended messages from his son, received after his son's death, are uniformly dark. For example, Mike's nickname Doc emerges as coined when Mike intentionally caused pain to an Iraqi prisoner by pressing on his wounds—to general laughter from Mike's squad mates. The temporal gap between these scenes and Mike's father's viewing of them, together with the terrible absolution offered Mike in his horrific death at the hands of his friends, casts the evidence of the videos in a difficult-to-assimilate light. We are left by the film with no answers but instead a kind of general distress. Mike's father concludes the film by flying a US flag—a treasured battlefield memento he himself carried in Vietnam—upside down, recalling an episode early in the film when he described such a sight as "an international distress signal" meaning "we're in a whole lot a trouble so come and save our ass 'cause we don't have a prayer in hell of saving ourselves." (The contrast with the neatly folded family flag at the close of *Act of Valor* [2012] exactly measures the distance between the first and second GWOT cycles.)

Zero Dark Thirty (2012) opens with a montage of phone calls to 911 on 9/11. Here, the familiar safety net of civilian emergency services is heard from a position of double helplessness: the audience knows what happened to the people heard on the tapes and knows that, on that day, there was little chance to save them. These civilian voices open a close point of identification for the film audience, enhanced by the startling use of intensely private materials for a public purpose. The American viewer is encouraged to imagine 9/11—images of which screenwriter Mark Boal felt were overly familiar—anew and in solely sonic terms. Following these overheard tapes, *Zero Dark Thirty*'s image track opens abruptly on a CIA black site where a detainee is being tortured.

In all of the above cases, recorded sound technologies bring disembodied voices into a film to thematically crucial effect. These are not moments that drive the plot so much as uses of dialogue heard by tape delay to establish a larger point about given wars and/or insert surprising voices into the mix—often voices gendered female. The target audience in more than a few cases is the movie audience as much or more than the characters within the film narrative. In this fashion, PCF writers and directors have shaped the ideological and expressive content of their films by exploiting the sonic technologies of modern war and life.[13]

VOICE-OVERS, ASSIGNED AND NOT

The film sound theorist Michel Chion describes the conventional movie voice-over in technical terms as a soundtrack mix featuring close miking and the absence of reverb. This is the sonic space Claire's initially diegetic voice moves into in *Hamburger Hill*. Chion dubs this the I-voice, and argues that it has the capacity to "function as a *pivot of identification*, resonating in us as if it were our own voice, like a voice in the first person."[14] One sort of I-voice, Chion writes, has "a certain neutrality of timbre and accent, associated with a certain ingratiating discretion. Precisely so that each spectator can make it his own, the voice must work toward being a *written text that speaks* with the impersonality of the printed page."[15] This theorization assumes that a neutral timbre and accent is possible. In fact, it is not. All voices suggest gender or race or class—especially the supposedly "neutral" voice of authority, in Hollywood typically mature, male, regionally unmarked, and white. This voice—often heard in pre-Vietnam combat films—is pointedly never heard in the PCF. Instead, PCF narration is assigned to select men whose lives are at stake in the story being told: Alvin in *The Boys in Company C* (1978), Willard in *Apocalypse Now*, Chris in *Platoon*, Joker in *Full Metal Jacket* (1987), Kovic just once at the very start of *Born on the Fourth of July*, Swofford in *Jarhead*, a different paratrooper in most every episode of *Band of Brothers*, the reporter Galloway in *We Were Soldiers*, Bob Leckie in *The Pacific* (2010), and, in a rare exception to soldier-narrators, James Bradley (son of the flag raiser John Bradley) in *Flags of Our Fathers*. All these voices anchor the viewer with particular men within the film story—even as each voice is anchored in a given character's body.

The Thin Red Line offers an extreme case of many voice-overs—some assigned to specific characters, others not. In Chion's terms, some of these acousmatic voices are de-acousmatized, others never are. Indeed, in the film's opening moments, writer-director Terrence Malick sets up this dual approach. The film's score, a seamless combining of compiled classical music and original cues by Hans Zimmer, plays an important role in establishing and supporting Malick's elaborate use of voice-over. Without this superabundant score, it is doubtful that Malick's conceit of male inner voices speaking their thoughts in voice-over would work.

Jungle sounds—birds prominent—come in on the title card naming the film. A low pedal tone enters, then an organ chord swells as the first image appears: a crocodile, framed in medium close-up sliding into scummy green water. Music takes over the mix, the chord sustained for as long as the croc remains above water—we and it float for an indeterminate period, a span impossible to measure or to anticipate ending. When the croc goes under, the organ chord releases as well. This point of synchronization between sound and image announces the film's stylized approach. What Dana Polan calls its "writerliness" could also be understood as its musicality, defined by the prominent placement of music in the mix and the explicit linking of music and image.[16] Many stretches of *The Thin Red Line* feature an image track following the music, even if, in this case, the release of the long organ chord, drawn from Arvo Pärt's *Annum per Annum*, cannot be anticipated by the viewer.

On the film's third image—a shot upward through the jungle canopy toward the sun, recalling images from early in *Platoon*—the film's first voice-over begins. A young-sounding man with a pronounced Southern accent asks, "What's this war in the heart of nature? Why does nature vie with itself? The land contend with the sea?" This voice returns four more times and, indeed, speaks the film's "last not-quite attributable words": "all things shining."[17] This voice is never assigned to any individual in the film, although one might have suspicions about which actor is speaking the words. One reading would place this rather defined voice—placed as to gender, region, and age—as that of the film's narrator. If understood in this way, Malick trades the articulate, mature, regionally transparent voice of documentary authority—as heard in *Guadalcanal Diary* (1943), for example—for a highly specific voice that carries instead an aura of authenticity (even as it is clearly structured around a stereotypical regional identity). In Chion's phrase, the unassigned Southern voice sits stubbornly *"at once inside and outside*, seeking a place to settle."[18]

Two musical cues accompany the Southern voice's initial voice-over. The first, a minute long, is by Zimmer. The second, two minutes in duration, is cut from the opening movement of Gabriel Faure's *Requiem*, titled "In Paradisum." Zimmer's cue provides an introduction to the Faure. The Faure moves toward the score's first unambiguous musical arrival—a satisfying cadence that lands at the moment Witt, played by Jim Caviezel, first appears. He is the first American seen in the film. As with the croc, Malick cuts Witt's entrance to preexisting music, shaping the temporal extension of *The Thin Red Line* to musical shapes drawn from compiled sources. Only in this case the harmonic motion of Faure's music guides the ear: Witt's appearance is prepared in the score, the first signal that he will be a source of stability, a character able to put the chaos of war into some sort of order.

Upon Witt's entrance, Malick turns to straightforward Hollywood techniques to fuse into one Witt's body, voice, and character of mind. His voice is heard first as voice-over, in a way that sets it apart from the Southern narrator already heard. Using eyeline matches, Witt is shown looking toward a native mother and child and heard speaking of his own mother's death. As the sequence unfolds, his words turn out to not be voice-over but a prelapped conversation with the other American soldier on the island. Malick cuts to Witt talking mid-sentence, just before he says, "I heard people talk about immortality, but I ain't seen it." A slow cross-fade from Witt speaking to Witt's memory of his mother dying follows. On the cross-fade back to Witt, pictured staring out toward the sea, his voice in voice-over lays out what might be called Witt's theme: the question, an urgent one in a combat film, of the proper way to die. "I wondered how it'd be when I died. What it'd be like to know that this breath was the last one you was ever gonna draw. I just hope I can meet it the same way she did—with the same [long pause] calm. 'Cause that's where it's hidden—the immortality I hadn't seen."

These two initial voice-overs—the first never de-acousmatized, the second securely assigned—typify the film to come. The unassigned Southern voice comes and goes. Its position relative to the diegesis is never resolved. Its questions—always heard to the sound of music—function as philosophical musings, lyrical reflections only indirectly related to the narrative events of the film. (On only one other occasion does an unassigned voice ask similar questions [see chapter 9].) A separate set of voice-overs are tied to distinctive individuals caught in the film's combat narrative: to Witt, but also to Private Bell and Lieutenant Colonel Tall. These voice-overs are fairly conventional, if always working in their own manner. Bell, for instance, speaks throughout to his wife. In the original script, Bell's were the only voice-overs in the film, imagined as letters—on the model of *Platoon* and many other PCFs, as discussed in the next section.

LETTERS READ ALOUD

Many PCF I-voices speak from a written text, usually one suffused with personal emotion: a letter or journal. Most notably, Chris in *Platoon* speaks aloud letters to his grandmother. (Oliver Stone wrote letters to his grandmother while in Vietnam.) With various levels of formality, these men speak to someone specific: we are allowed into their story for having chosen to watch their film. As Alvin says in *The Boys in Company C*, he might someday publish his journal—the understood text of the voice-over—"that is, if anybody even wants to read about what happens to a bunch of guys they're gonna send to Vietnam." In effect, Alvin compliments the 1978 film audience for having bought a ticket to a film about Vietnam.

To very different ends, *Saving Private Ryan* and *Letters from Iwo Jima* (2006) gather into overlapping sonic montages the sounds of letters read aloud by their authors, who are not characters in the film story. Brief analyses of these parallel sequences furthers previous discussions of acousmatic voices as moments to dwell rather abstractly on combat questions and as a kind of word-made music, open—as music is—to different interpretations.

Official letters from officers to the families of men going into battle or killed in combat are a special feature of the subgenre's military setting. Written and spoken by men of authority, such letters carry a kind of formality but are, nonetheless, personal. They articulate the intersection of the military and the domestic, just as the arrival of uniformed soldiers on Casualty Notification duty does in visual terms in several films. Condolence letters sent by officers to the families of the fallen can speak to individual losses even when those losses are only registered in passing in the film. In a brief but powerful passage in *Saving Private Ryan*, the overlapping voices of officers speak the texts of condolence letters being typed at the War Department in Washington, DC. In the original script, Rodat imagined a cacophonous sound match between "The SOUNDS OF BIG GUNS and MACHINE GUN FIRE" and "CLATTERING TYPEWRITERS."[19] In the film, Spielberg smooths the transition by musical means—John Williams's score plays throughout—and adds the anonymous officers' voices, which have the effect of putting off the unfolding of the plot. The first hint of the film's story—about a half hour in—comes at the end of a long crane shot floating above the dead GIs on Omaha Beach, which ends on the back of a Ryan brother. The viewer is primed to expect the plot of the film to begin.

But the cut to the War Department undermines this desire: Spielberg pulls away from the specifics of the film story and dwells briefly on the avalanche of grief created by the landings. Overlapping officer voice-overs, mixed to make specific phrases comprehensible, personalize this grief, even as the clattering typewriters suggest the scale of the task of dealing with what Rodat calls "the paperwork of death." (Chapter 10 returns to this moment to consider the role of elegiac music in the mix.)

Letters from Iwo Jima is, even in its title, obsessed with letters. The doomed Japanese soldiers on Iwo Jima write letters home beyond the point when they can be sent off the island. Late in the film, Saigo, the youngest of the main characters, chooses to bury the unsent letters rather than destroy them as ordered. The film's framing scenes—set in the present day, analogous to the old Ryan scenes in *Saving Private Ryan*—show archaeologists finding the bag of letters Saigo chose to save. In the film's closing moments, these letters spill out of the bag in slow motion while the overlapping voices of their dead writers are heard in a collage of untranslated voice-overs. The effect is similar to the condolence letters in *Saving Private Ryan*—except for a non-Japanese-speaking audience, the content of the letters is opaque. Lars-Martin Sørensen has argued that this film successfully welcomes a variety of perspectives on World War II for its Japanese audiences. While the character of Saigo—who thinks of personal survival over national pride—will not please pro-military viewers, Sørensen notes, "The nationalist pathos is still suitably thick to please the hardliners." Specifically the overlapping voices at the close are "easily construed as the sound of the spirits of the fallen ... *gunshin*—'soldier gods'—or *kami*: Shinto spirits that are enshrined and commemorated at the Yasukuni shrine [in central Tokyo] ... whether Eastwood intended it or not."[20] Voice-overs here lie open to specific cultural readings that cut against any perceived antiwar message of the film. The honorable words said by the officers in *Saving Private Ryan* can, of course, be read in a similar manner if without the formal nationalist militarist religion of Japan evoked for some in *Letters from Iwo Jima*.

Letters heard as voice-overs also provide a way for the dead to speak—opening a crucial space where the fallen might live again. As Chion notes, "What could be more natural in a film than a dead person continuing to speak as a bodiless voice, wandering about the surface of the screen?"[21] And what could be more natural for a genre obsessed with properly memorializing the dead than for the voices of fallen soldiers to be heard in this way. Surprisingly, letters spoken in voice-over by the dead was a device only added to the genre in the mid-1990s.

In *Courage Under Fire* (1996), Karen Walden, the chopper pilot played by Meg Ryan at the center of the *Rashomon*-style narrative, speaks aloud the letter she left behind for her parents. (Recovering the letter and delivering it to her parents forms part of the plot.) It's the film's only representation of Walden on her own terms: she is otherwise presented in the memories of the men she commanded—memories shaped into lies at times—or moments imagined by Lieutenant Colonel Serling. Only in the letter does Karen speak for herself. And she does so in utterly colloquial terms, enhanced by her soft Southern accent and even, confident, gentle tone.

> Dear Mom and Dad, Well, this is it—the big push. Looks like it's really gonna happen. And I'm afraid. Not of being hurt or killed—well, kinda but not much. My only regret will be to never see you two again. And that I'll never see Anne Marie grow up. But I know she's in good hands. The best. What I'm really afraid of is that I might let my people down, my crew. These people depend on me. They put their lives in my hands. I just can't fail 'em. Now I know if you get this letter it means I'm dead. I only hope that I made you proud and that I did my job and I didn't let down my country, my crew, and my fellow soldiers. I love you guys. Never stop telling Anne Marie how much I love her. Your daughter, Karen.

The ironies of this letter are hard to process while watching the film. In fact, Karen's crew lets *her* down, leaving her to die in a huge, fiery burst of napalm. But the situational complexities of the letter are washed away in the commonplace valor of Karen's words and voice. There is a lack of posturing in her text and delivery, aided by her gender, of course, but not inaccessible for male warriors as well. (*Black Hawk Down* closes with a very similar voice-over letter from the dead spoken in tender tones by a man.) By this point in the film—Karen's letter is the film's final words—*Courage Under Fire* has turned toward the enactment of rituals of resolution and homecoming. Yet, in the ambiguous way of many PCFs, such resolution always involves the act of dwelling on the fallen, remembering the lost. This affective move was necessitated by the 1980s Vietnam cycle's burden of rehabilitating the memory of the men who fought a lost war. *Courage Under Fire* demonstrates the efficacy of such sentiments applied to a war the United States "won," showing for the first time the flexibility of the PCF's originating expressive goals.

The SEAL PCF cycle includes two letters from the dead heard as voice-over. The letter in *American Sniper*—a film with a relentlessly diegetic soundtrack—only implies voice-over. In fact, the voice reading the letter is prelapped: we hear the start of a scene before we see it. An older woman's voice—one of a handful of women who speak in the film—breaks into the soundtrack. Her voice enters on images of Kyle and two comrades riding home in the back of a C-130, keeping the requisite constant watch over five flag-draped caskets, a SEAL called by the nickname Preacher's among them. The voice says, "Glory. Something some men chase and others find themselves stumbling upon, not expecting to find it. Either way it is a noble gesture that one finds bestowed upon them." These troubling words—presented in what is for this film a startling mix—nudge the viewer, briefly, to think a bit about Kyle. Which sort of hero is he? The one who chases glory, or the one who finds it unexpectedly? In solid PCF fashion, no easy answer is given. Instead, a variety of viewer responses are opened and left for the individual to do with as they will. On a cut to Preacher's funeral, the speaker is revealed (or implied) to be his mother. Here the voice-over turns diegetic as the letter turns toward a questioning of glory, a move that leads to a slight breakdown of grammar: "My question is when does glory fade away and become a wrongful crusade or an unjustified means by which consumes one completely? I've seen war and I've seen death." A sharp, unseen male voice interrupts the mother (and her son) and initiates a deafening twenty-one-gun salute, to which Kyle's wife Taya visibly shudders while Kyle remains literally physically unmoved.

The text of this letter is authentic if incomplete. The SEAL Marc Lee penned it to his larger group of family and friends in 2006 as a July 4 greeting, shortly before his death, which is shown graphically in the film and which Kyle takes very hard. The full letter—more than eight hundred words in length, available on the Internet—moves from its initial discussion of glory through a critique of American popular culture toward an affirmation of private values, concluding, "I think the truth of our [nation's] greatness is each other. Purity, morals and kindness, passed down to each generation through example."[22] As used in the film, Lee's complicated argument is severely truncated, suggesting only doubt at the value of fighting and implied critique of US actions in Iraq. While insisting that they supported *American Sniper*, Lee's wife and mother both used conservative media outlets to argue that Lee was devoted to his SEAL brothers and the US mission in Iraq.[23] By quoting selectively from Lee's letter in a striking use of pre-lapping, the film employs an authentic text to reposition at least one SEAL as a skeptic, if not a critic, of the Iraq War, giving viewers along that part of the ideological spectrum a nod and keeping *American Sniper* from becoming a completely gung-ho military film. In this case, moderating or editing a soldier's language kept a PCF from offending what might be termed the antiwar contingent of the audience.

Preacher's letter threatens to take *American Sniper* toward an interpretation of the Iraq War as a hubristic exercise in glory-seeking violence. It's a troubling eruption of doubt into the film, intensified by its introduction into the formal space of a military funeral, a ritual zone the SEAL films treat with great care (and typically much music). Put into the words of a grieving mother, the moment chimes with Clean's mother's tape in *Apocalypse Now* and Claire's in *Hamburger Hill*. These moments, varied in content as they are, use voice-over to bring women into the PCF, their startling, sudden sonic presences activating for the film audience a connection to the world of battle and its consequences. For female viewers, these voices are among the only points of direct identification the genre offers.

No words of doubt about the value of soldierly sacrifice mar the military funeral at the close of *Act of Valor*. That's not to say that the moment comes off without intense sadness. If anything marks *Act of Valor* as more than a pro-military action movie, it's the time spent on the LT's funeral and the funeral's placement in the larger span of the narrative. The 1990 action flick *Navy SEALs* also includes the death of a SEAL and a funeral, but in the middle of the narrative—not at the close. This allows *Navy SEALs* to touch on ultimate sacrifice and then, in the thrill of an extended urban warfare sequence, forget it in a final blaze of action. *Act of Valor's* final action sequence, a chase on foot through a tunnel between the Mexico–US border, ends with the wounding of the Chief, who is seen in a wheelchair at the funeral, and the sacrificial death of the LT, who falls on a grenade. (The sequence of musical events at the close of the film are detailed in the final chapter.) Here, it's important to consider how a film-long strategy of ambiguous voice-over narration presented as a letter concludes in a revelation of who the letter is from and to whom it is addressed.

Act of Valor opens with a voice-over offering advice to an initially undisclosed listener. Clues are dropped early on in reference to "your father." The recipient of the letter, it turns out, is the son of the LT (the SEAL who dies), the son whose impending birth is discussed across the story and whose birth the LT misses in the line of duty. The author of the letter is the Chief, standing in for his dead friend. At the close of the film, during a passage from the letter quoting a poem by the Native American warrior chief Tecumseh, the LT's voice sneaks in, speaks together with the Chief, then takes over entirely. As in *Courage Under Fire* and *Black Hawk Down*, the dead speaking a letter forms *Act of Valor's* final words. This handing off between men sounds over the image of the LT's wife reading the letter and his infant son sitting nearby in a high chair. For some young male audiences—in particular the elite private university students in my classes with whom I viewed *Act of Valor*—the pressure the film places on the little boy to grow up to be a man in the mold of his SEAL father is tremendous and to be resisted. Granted, male students at a different university would likely respond very differently.

But *Act of Valor* is exceptional—and exceptionally manipulative—in this regard. (*Lone Survivor* is framed by classical voice-overs spoken by Luttrell over images of his battered body; see chapter 9.) Indeed, one mark of the GWOT cycles is an avoidance of voice-overs that are not practical— either prelapped voices or informative texts read aloud. The only voice-overs in *Zero Dark Thirty* and *Green Zone* (2010) are emails read aloud as a means to deliver text in an age of electronic messaging. *The Hurt Locker*, *The Kingdom* (2007), and *United 93* (2006) lack voice-overs of any sort. The sonic device of the disembodied voice, used regularly in earlier cycles and consistently associated with access to the inner thoughts of American soldiers, largely dries up in the twenty-first century, leaving viewers with more work to do to interpret these films and the soldiers represented in them.

PART III

Sound Effects

6. Nothing Sounds Like an M-16

War is a catalogue of sounds. Our ears direct our feet.
—GUSTAV HASFORD, *The Short-Timers* (1979)

Inexperienced soldiers learn quickly to be good listeners. Interpreting how sounds define the battle zone is a basic first step for any cherry or FNG (fucking new guy). Early in *Hamburger Hill* (1987), the nervous Bienstock cries out about artillery rounds, "I can't tell the difference between incoming and outgoing." In episode 9 of *The Pacific* (2010), Sledge, now a combat veteran, mocks replacements joining his squad on Okinawa who are similarly ignorant of such basic battlefield signals.

Combat movie audiences also need to be instructed in how to follow battles by ear. Teaching a cherry to interpret sounds also teaches the moviegoer. In the script for *The Thin Red Line* (1998), a character identified only as YOUNG SOLDIER asks Witt, "I never been shot at before. Is this what it sounds like?" Ever calm, Witt replies, "This is what it sounds like. We're bein' shot at."[1] In *Black Hawk Down* (2001), Grimes asks Waddell in the nick of time how to hear the difference between gunfire that's aimed and gunfire that's not.

> WADDELL: "A hiss means it's close. A snap . . ."
>
> SNAP!SNAP!SNAP!
>
> WADDELL: "Now they're shooting at us!"[2]

Occasionally screenwriters spell out the sounds of combat step by step, helping the viewer make sense of an unfolding battle. *Casualties of War* (1989) opens with a group of soldiers on nighttime jungle patrol. The battle begins acousmatically, with explosions in the distance. The CO says twice, "Fix that sound." Another soldier names the weapon heard: "I hate fuckin' mortars, goddamn it." The RTO (radiotelephone officer) reports that a nearby company is getting hit. One soldier asks, "Shit, man, do they know we're here?" Another works the map, trying to "fix that sound." The CO asks the RTO to be ready to call in an artillery strike on the "tubes"—untranslated lingo for mortars—after which, he notes, "We'll adjust by sound." Then, an inexperienced soldier fires his automatic rifle at a threat in the dark that only he sees. This dangerous lapse of noise discipline, followed by angry shouts from the other soldiers, invites greater precision from the enemy's mortars. Very soon the first seen mortar arrives. The enemy has found them: the CO confirms the obvious, "The tube got a fix on us." Up to this point, this unit has been fighting by ear. *Casualties of War's* opening scene lays out a jungle battlefield with geometric precision while also providing just enough verbal information to cue the viewer to how the soldiers are interpreting and reacting to events. Sound effects and dialogue set the stage for the arrival of seen explosions, which kick off the next, much noisier, more dangerous phase of the battle. J. Martin Daughtry's "concentric zones of wartime audition," a model of soldierly listening gleaned from interviews with veterans of Iraq, nicely maps onto the above scene, with the battle moving from the narrational zone, where soldiers attend to "the unique sounds and approximate locations of particular weapons" ("fix that sound"), to the tactical zone, where soldiers use "their skills of echolocation to determine the proximity of explosions" ("adjust by sound"), to the traumatic zone, where "these sounds lose their capacity to serve as a resource, a text to be interpreted," and become immediate threats to survival.[3]

Experienced soldiers read battles in progress by distinguishing the sounds of different weapons. Captain Robert Hemphill—who commanded Oliver Stone in Vietnam and remembered him as a good soldier—fills his 1998 memoir *Platoon: Bravo Company* with a consistent vocabulary of poetic approximations of weapon sounds: grenades *"Crump!"* and rockets *"Whoosh!"*; American M-16s go "rat-atat-tat" while the enemy's AK-47s go *"Pop-pop-pop,"* thankfully an "easily recognizable sound." Hemphill recalls the confusion of a battle where "the gooks seem to be using an [American] M-60 or something that sounds like one."[4]

Different wars have different weapons—in effect different instruments, producing different sounds and yielding different sound-told war stories. Iraq veteran Brian Turner's poem "What Every Soldier Should Know" offers sound advice: "You will hear the RPG coming for you."[5] This line applies nicely to *Black Hawk Down*, its noisy battle zone crisscrossed by this weapon that moves through space along an audible line of fire. Different weapons speak different rhythmic languages, allowing for a kind of counterpoint: as Gustav Hasford writes in *The Short-Timers* (1979), "The crack of a sniper's rifle pierces the muted rhythm of Mother's machine gun."[6] Brass shell casings falling from a minigun cascade like chimes in *Black Hawk*

Down and *Act of Valor* (2012), and in *Zero Dark Thirty* (2012) serve as shining echoes of precisely placed kill confirmation shots. A single casing hitting the floor with a resonant ring gives away an entire unit's position in *American Sniper* (2014). Indeed, the characteristic sound of a weapon distinguished in the sonic chaos of battle has been used as the hinge for an entire PCF plot. Evidence for Karen Walden's final moments in *Courage Under Fire* (1996) come down to the sound of an M-16, known to be in her hands, heard firing just before US helos drop napalm on Walden's last known position. As an ear-witness soldier testifies, "Nothing sounds like an M-16."

Few weapons are as rhythmically fitted to narrative cinema as the grenade, the use of which punctuates combat action with a pregnant pause between the tiny click of the pin being pulled and the loud explosion that follows seconds later. We hear both click and boom when Sergeant Keck, played by Woody Harrelson, dies in *The Thin Red Line* after having inadvertently pulled the pin while grabbing a grenade from his belt—a "fucking recruit's trick" he shouts. Keck muffles the explosion—and protects his comrades—by falling against the grenade, absorbing its blast. The time between click and boom opens a moral space within which soldiers make character-defining choices. At the end of *Act of Valor*, this span of time is elongated by slow motion and, unusually, a slowed-down cry of "grenade" from the LT, who soaks up the weapon's force with his brawny, gear-heavy body, which is thrown into the air. No such dramatic heightening of the time between click and boom is granted Alvin at the end of *The Boys in Company C* (1978), when the slightly built soldier falls on a grenade to protect some nearby children. Alvin's act of valor is presented matter-of-factly rather than mythologized. No representation of the time between a grenade's click and boom carries the moral horror of the suicides in *Letters from Iwo Jima* (2006). Japanese grenades are activated by hitting one end against a hard surface, an act of bold intention unlikely to be accidental. One by one, a group of Japanese soldiers follow their officer's lead, slamming a grenade against their helmeted heads then hugging the explosion, killing themselves by ripping open their chests. In *Flags of Our Fathers* (2006) the sounds of these deaths are heard from outside the cave by uncomprehending American soldiers, who then enter the cave and view the blasted bodies.

An ideal weapon for the continuous-action film, the RPG (rocket-propelled grenade) is, in essence, a bullet visible in real time that moves along a clear progression of sonic events: a small explosion on firing (always associated with intentional human action), a whoosh while in flight, and a big explosion on the easily anticipated moment of impact. Tracing a line in space, time, and sound, the RPG is a melodic weapon-instrument. *Black*

Hawk Down plays variations with the RPG, using it to mark out the space of Mogadishu, both at street level and in the skies (RPGs take out both downed Black Hawks). Ken Nolan's script uses action words recalling cartoon captions to mark the sound of RPGs in flight: "WHOOSH!," "a loud FWOOSH!," "BAWHOOM!," and "PHWOMP."[7] Perhaps the most surprising PCF RPG is the one that downs a Chinook full of SEALs coming to the rescue of their embattled brothers in *Lone Survivor* (2013). We neither see who fires it nor hear it being launched. Its whooshing sound is not privileged in the mix, which instead offers emotional misdirection, a rising drum cadence in the score signaling imminent rescue. Only panicked cries of "RPG" within the helo alert the viewer to the incoming ordnance. And so, when the RPG enters the Chinook's open bay and explodes to down the helo—killing all on board—the viewer is pulled up short: the rising martial music abruptly cut off by a blast that comes out of nowhere from a weapon typically associated with the cinematic pleasure of seeing and hearing a complete life cycle from shot to explosion.

Combat movies and actual combat alike offer intense experiences of a particular kind. As one veteran of the invasion of Iraq observed: "I mean it was a front seat to the greatest movie I've ever seen in my life. . . . I loved it. Anytime I get shot at in a firefight, it's the sexiest feeling there is."[8] In his 2010 book *War*, Sebastian Junger lays out the trade-off inherent in living on the front lines: "The one absolute impossibility at Restrepo [a remote firebase in Afghanistan]—you could even get booze if you wanted—was sex with a woman, and the one absolute impossibility back home was combat. Whether the men realized it or not, they had made a rough trade where one risked becoming a stand-in for the other."[9] Choosing to watch a war film—almost always with an all-male cast, with graphic violence but no graphic sex—forces a similar trade-off. Indeed, combat scenes are the sex scenes of the war film genre.

As the war film scholar James Chapman notes, combat scenes by definition require "the imposition of structure onto the battlefield."[10] War rendered on film or video becomes a fixed and framed activity in a way actual combat can never be. So while the US Army regularly creates a post-battle "event storyboard" from which "a kind of clarity can emerge," reverse engineering a timeline of the action in a quest for "lessons learned," action filmmakers storyboard their simulated battles *before* shooting, for the sake of efficiency and safety during production and so that a specific tale might be pieced together in the editing suite and a precise point drawn by the moviegoer.[11] This ability to take the chaos and uncertainty out of war—to "plan" a battle and control who lives or dies—opens all war films to the understandable criticism of never getting close to the reality of combat. (Gulf War commander General Norman Schwarzkopf used a musical analogy to describe the futility of trying to plan the course of battle: "It is choreographed, and what happens is the orchestra starts playing and some son of a bitch climbs out of the orchestra pit with a bayonet and starts chasing you around the stage. And the choreography goes right out the window.")[12] But storyboards and the films made from them can just as easily frame combat action as chaos as make it unrealistically legible. (Schwarzkopf's extended orchestra concert metaphor suggests many a PCF—with, for instance, the Chinook-downing RPG in *Lone Survivor* cast as the son of a bitch with a bayonet.) In short, filmmakers have the option to present combat as chaos or not. Both options entail a clutch of aesthetic and technical choices.

Still, PCF makers must tell coherent stories—these are commercial narrative films, after all. PCF battle scenes tend to seek balance among competing possible aesthetics for the presentation of combat: representing the pure sensation of battle (as thrill or terror); staging a meaningful order of battlefield events (victory, defeat, or stalemate); and/or putting forth an ideological or philosophical understanding of the modern battlefield, a space open to topics such as the properly masculine response to mortal threats; the nature of the enemy; moral action in a zone of sanctioned killing; human frailty and failure; heroic action and sacrifice; the powerful bonds of brotherhood; the ironies of chance versus choice in human life; the very efficacy of war. PCFs—action films in the end—offer sensation but also present combat as a richly significant human experience. Sound—especially music, with its openness to varied interpretations—plays an essential role in making combat meaningful.

This chapter opens a multi-chapter consideration of sound in combat sequences. Below, battle sequences dominated by sound effects are discussed in detail. Given the importance of music in the PCF, it comes as no surprise that many combat set pieces include music. This chapter also includes effects-driven combat soundscapes that selectively incorporate music for precise ends. Subsequent chapters (7, 8, and 9) treat more thoroughly musicalized battles within their respective larger topics (helicopters, unmetered musics, and metered musics). The primary goal for all these battle scene analyses is to understand how the soundtrack helps shape combat set pieces into meaningful wholes functioning as narrative units within larger films. Such set pieces must work simultaneously toward various ends. Plot-wise, they chart the progress or nature of a given skirmish, battle, or war. Toward the goal of character development, they display soldiers in their element as men of action (or not). Finally, combat set pieces can also contain or put forth the core meaning of a film, allowing actions—often as defined by sounds or music—to speak in place of words.

Advances in cinema sound have consistently expanded the sonic possibilities for PCFs, opening the door toward immersive and expressive combat soundscapes. Dolby arrived in the late 1970s, just as the subgenre was emerging. As Pauline Kael noted of *The Deer Hunter* (1978): "The superlative mix of the Dolby sound gives a sense of scale to the crowd noises and the voices and the music; we feel we're hearing a whole world."[13] Dolby brought increased range and clarity and the addition of a speaker at the back of the theater. These changes to cinema sound made the sonic experience of the Vietnam cycle of PCFs qualitatively and quantitatively different from all previous war films—automatically making such films more immersive for original viewers. As discussed in chapter 2, *Apocalypse Now* (1979) in its premiere 70mm showings used a Quintaphonic mix explicitly drawing on rock-concert aesthetics—part of the film's conceit that rock and the American way of war in Vietnam were fundamentally similar. The introduction of digital surround sound (DSS) in the mid-1990s—which divided the back surrounds into right and left and further enhanced film sound clarity and dynamic range—came just before PCF makers turned toward wars other than Vietnam. *Courage Under Fire* (1996) was the first PCF to use digital, and *Saving Private Ryan* (1998) spectacularly exploited digital sound—previously associated with fantastical effects—to put the audience into the D-Day sequence with immersive power. Part of the latter film's overwhelming effect in the theater was the novel impact of these new sonic resources. By the time of *Black Hawk Down* in 2001, critics were complaining about sensory overload and drawing connections to other media formats. The critic John Powers wrote, "Filmmakers now have the technical ability to plunge us into the very bowels of battle—rattling our seats with explosions in Dolby Digital, rattling our sense with jittery camera work and tracer bullets whizzing by our ears. They've turned war into an Extreme Game."[14] Powers highlights volume and spatial placement—tools for increased immersion, as in video games. Immersion in the battle space—"bullets whizzing" from the speakers behind the screen to the surrounds or the reverse—remained part of the genre.

American Sniper (2014) is the first PCF mixed in Dolby Atmos, a system that allows for precise placement of sounds in the surround by way of multiple speakers distributed grid-like across the walls and ceiling of a theater. Few viewers saw *American Sniper* in an Atmos theater, but even the film's

Dolby Digital mix endeavored to create a precision sort of sonic immersion. As noted, in *American Sniper* the mix virtually always suggests the position of the camera. For example, a quiet shooting-range scene was shot and cut using shifts of angle from behind Kyle (he shoots away from the camera) and toward Kyle (he shoots toward the camera). The mix for each bullet is different. Shots framed from behind Kyle, looking downrange at the target, make a sound that stays behind the screen. Shots going toward the camera, framing Kyle from the front, whiz past the audience and hit one side of the back wall of the theater. In an Atmos mix, these shots hit a precise spot on the back wall or ceiling. Given the deliberate, isolated pace of Kyle's shooting and the relative quiet of the scene, the effect of Kyle's shots "into" the theater is overliteral and exaggerated. (Scenes of actual combat are seldom so clear cut.) The mix for the shooting range exposes how effects and the mix work to either spell out the implied sonic space of a scene or instead accompany the image track in a more evocative manner. The shot that travels through the theater—past/over/through the viewer, who might even duck—activates an odd immersive position: no one stands between shooter and target at a shooting range. By contrast, the shots that stay in the world of the film do not activate a sense of spatial placement in the scene, instead keeping the viewer's attention on the theme of this sequence: Kyle's aim compromised by his new relationship with Taya. We are invited to reflect on—or, perhaps, chuckle at—how a man might be altered by a woman. Sound often plays a key role in directing the viewer toward immersion in combat action or reflection on combat and its larger meanings. Sound choices cueing reflection prove an important resource for PCF action scenes, making room for combat to *mean* in a more general sense for the viewer, who is enjoined to both follow the plot—who lives or dies in the story—and also think more broadly about the experiences of real American soldiers in similar, but real and actual, situations.

The eleven combat sequences in *The Pacific* employ a range of options for activating immersion and reflection.[15] The great variety of the series springs from its three central characters, each of whom brings his own perspective to the experience of combat. Robert Leckie, John Basilone, and Eugene Sledge's respective combat experiences on Guadalcanal, Cape Gloucester, Peleliu, Iwo Jima, and Okinawa, as represented in the series, use sound to shape the viewer's understanding of combat. The vast majority of combat in *The Pacific* relies on sound effects to activate different viewer relationships to the action: whether immersion (in tactics or risk) or more detached reflection on the nature of the Pacific War. A few sequences use music in targeted ways. Below, I consider how the combat scenes in *The*

Pacific lay out the options open to mixing a battle with only or mostly effects. Parallel examples from other PCFs, including iconic battles from the World War II and GWOT cycles, are analyzed along the way, shedding light on how different PCF cycles about different wars have figured the soldier's experience of combat in sound.

Episodes 1 and 2 of *The Pacific* each include a combat sequence set on the island of Guadalcanal. Both involve nighttime frontal attacks by masses of Japanese soldiers on entrenched US positions. In both cases, the Americans hold the line with heavy automatic weapons fire. Fields of Japanese dead and wounded are revealed in the light of day. The first sequence—Alligator Creek, coming just thirty-five minutes into episode 1—features Leckie and his group. Combat action consists of open and continuous firing across a wide creek lasting without cease for three and a half minutes of screen time. There is little narrative content beyond the start and finish of the firing: the battle is a matter of sheer duration. Individual deaths are not highlighted, nor are any individual acts of combat (beyond two vague changes of position that are not located within any clear battle space). The only substantial event is the terrified reaction of an American officer, who cowers in fear. All the rest of the Americans do their job. It's mostly just firing: a "turkey shoot," Leckie says after it's over. The image track is a chaotic montage of flashes and firing in the dark across the creek. Sustained fire, not all that precisely synchronized with on-screen events—with the exception of one mortar—fills the soundtrack with unremitting noise. Given the domestic context for watching a cable television series, the viewer might choose to grab the remote and turn down the volume. (I did.) The battle concludes with a frontal shot of a machine gun firing directly at the camera, shown four times, each shot framed a bit closer, lasting a bit longer, sounding a bit louder, the final time leading to a whiteout that takes the narration into the next day. The rhythm and volume of the set piece comes to a peak, then cuts off abruptly. Throughout the viewer is enjoined to assess the terrifying danger of the combat context—interestingly, at the conclusion, from the enemy's position directly in front of an American machine gun. Reflection comes the next morning, when *The Pacific*'s mournful string-heavy score plays over the Japanese dead and the Americans looking on the scene. Leckie endures a similar if more terrifying night battle—in the rain, at closer range—on Cape Gloucester in episode 4. These two similar battles offer undifferentiated spans of terrible noisiness to be endured and, hopefully, survived.

By contrast, the other battle sequence on Guadalcanal uses editing and the effects track to narrate heroic action. In episode 2, John Basilone's Medal of Honor–worthy combat actions are presented in a way that locks the viewer into his decisions and actions without adopting the hero's point of view. Again, it's a night battle against advancing Japanese. But this time the image track and soundscape are oriented around a specific machine gun— Basilone's—which has a lower-pitched sound, different from others near him. On shots of Japanese falling to machine gun fire, we can hear that their deaths are coming from Basilone's weapon. In an interlude of hand-to-hand combat, Basilone's weapons again dominate the soundtrack, bursting powerfully and in synchronization with his actions against background battle noise that serves as so much accompaniment for the musical rhythm of his deeds. Sounds synced to Basilone's movements clarify when specific Japanese soldiers die at his hand. In the final phase of the battle, Basilone's sustained firing of his machine gun begins to attract the attention of the other American soldiers. They turn at the sound of his weapon, watching and listening as he fires. The sequence ends with Basilone firing his machine gun again, its sound again filling the soundtrack, including on an image of his hand on the trigger (a reverse image of the shots into a firing machine-gun muzzle ending the Alligator Creek sequence). Sound, applied with consistent audio perspective and amplifying a single soldier's actions, helps the viewer stay with Basilone and follow his reaction to an environment not so different from Leckie's, except that here meaningful direction and not simple duration defines the battle. Throughout, the soundtrack concretizes Basilone's acts sonically but always with a measure of restraint. For example there is no sonic reinforcement—no sounds of sizzling skin—when Basilone picks up a hot machine gun barrel, burning his lower arms.[16]

Basilone's other battle sequence—the Iwo Jima landing ending in episode 8—similarly tracks heroic intention. In a context of intense, withering fire reminiscent of several other amphibious landings, Basilone leads his group of young Marines with his voice: "You wanna live? Get off the beach," he yells. In shots from behind the running Basilone, his voice continues to be heard (unrealistically) above the battle, crying, "Keep moving." He moves with purpose and resolve in an impossibly chaotic situation, seemingly impervious to the danger, leading his inexperienced men to take the first line of Japanese defensive positions. His voice—that of the drill instructor—carries the soundtrack. But even Basilone succumbs, his voice gradually drowned in the chaos of the situation. Music enters the mix just prior to Basilone's fatal wounding: the score sneaks in and the noises of battle fall back just before bullets rip a bloody swath through the middle of his body. He falls and the mix immediately tilts toward music, moving quickly to near diegetic silence. The image track goes to slow motion. This shift in the soundtrack signals the viewer to abruptly shift their mode of engagement from following Basilone's tactics to reflecting on his death as it happens: Basilone's young Marines similarly pause in their firing to watch their leader die. His death is drawn out in a long musical cadence. The point of closure—when he succumbs—cuts abruptly to his now-still body framed from above on a sudden bump in the effects track. Having paused for the hero's death, the battle rages on as the episode ends. Such abrupt shifts from immersive action to reflection on loss are not unusual in combat set pieces and music proves a ready resource at such moments.

Cueing viewer reflection with effects only rather than music seems to require an image track that narrows the experience of battle down to directly subjective terms. In these cases, the mix plays an essential role in rendering discrete stretches of film unambiguously subjective. Basilone's Iwo Jima contrasts sharply with the other amphibious landing under heavy fire in *The Pacific*—Sledge at Peleliu in episode 5. The sequence begins with Sledge's ride to the beach in an LVT (Landing Vehicle Tracked). The LVT's emergence from the hold of a ship is dramatically set up both musically and visually: we pass with Sledge through a whiteout and musical crescendo into a CGI vista of a massive amphibious attack. The soundtrack from here on is entirely effects: the LVT's motor, planes overhead, incoming Japanese shells causing bursts of water, large-caliber machine guns on the boat—two minutes of noise. On arrival at the beach, the camera stays on Sledge. A mortar goes off near him and briefly the soundtrack shifts to subjective sound: Sledge's ears' ringing. (Sledge has entered Daughtry's "trauma zone," where "belliphonic sounds produce physiological damage," in this case temporary.)[17] Crucially there are no POV shots: we see Sledge and hear both what he hears (a high ringing tone) and the intimate sounds he makes (small gasps and panicked cries). On a camera move, the soundtrack slips cleanly out of Sledge's subjective sonic perspective: sound seems to motivate the move, with the camera turning away from him toward an offscreen cry of "corpsman." After this, Sledge is in the fight, able to move with intention and resolution.

The clear precursor to Sledge's landing is the D-Day sequence in *Saving Private Ryan*. In that set piece, shifts in the mix similarly suggest that the viewer is looking on the scene through a particular perspective. Twice the soundtrack cuts abruptly into Captain Miller's subjective sonic perspective. We see his (Tom Hanks's) face and hear what he hears. But we also, briefly, seem to see the devastating scene through Miller's eyes. The more mature Miller offers a point of reflection on the scene—he's an English teacher, after all—unavailable to the terrified young Sledge. As with Sledge, the proximate cause of Miller's combat deafness the first time is his being too near an exploding shell. This narrative conceit both explains his deafness and offers a kind of punctuation: a point of abrupt transition from one mix to another. Artillery shells prove useful this way in many PCFs. Sledge and Miller's episodes of subjective sound offer the viewer a respite from the volume of fire. In Miller's case, the viewer—together, it seems, with Miller—looks upon the scene with something resembling distance. Here, in the midst of the most immersive of PCF battles a muffling of diegetic sound offers us time to think about what we are seeing. Such reflection is essential to the PCF: we are supposed to contemplate and cannot do so if we are immersed in action all the time. Episodes of subjective sound seem to be the only way to do this without activating the score, a move that has the potential to pull us back too far (as subsequent scored battle scenes analyzed below will show). The effects track so deployed plays an essential meaning-making role in many PCF combat scenes.

The Peleliu landing in *The Pacific* moves inexorably inland with no sense of where the stopping point—the soldiers' place of rest—might be. By contrast, Steven Spielberg's D-Day landing in *Saving Private Ryan* establishes at the start in sonic and visual terms the ground to be covered on Omaha Beach. The geography of the actual battle aided Spielberg in telling his story. Americans taking Omaha Beach had first to find shelter at the shingle—a sandy shelf that offered natural protection—then move from the shingle to the hills above the Atlantic Wall, the German army's fortified line of defense. The film's edit and mix localizes our experience of this progress by way of consistent auditory perspective and a major shift in the ambient natural sounds of the battle. As Robert Burgoyne has noted, the first bullets seen in the D-Day sequence are heard only as impacts into flesh and not as shots fired.[18] Only part of the projectiles' life cycle is given—analogous to the RPG we don't hear or see fired in *Lone Survivor*. But immediately after these shots are fired, Spielberg cuts to a reverse angle, from the perspective of the German machine gun position on the Atlantic Wall: the distance to be covered is succinctly laid out.[19] On five more occasions before Miller and company reach the shingle, Spielberg cuts back to the view from the German machine gun. Passage to the shingle, the first ten minutes of the sequence, is supremely chaotic. It is also, however, finely structured. Spielberg mixed the sequence for variety, likely so as not to completely overwhelm the viewer. (Many were overwhelmed anyway.) For example, just after the firing starts, the camera plunges underwater. The immediate effect for the viewer is a muffling of the overall volume of the soundtrack—a relief (of sorts). Bullets sound different underwater and, somewhat perversely given the context, there is a kind of beauty to bodies suspended in water. (A similar moment in the Australian war film *Gallipoli* [1981]—men fired on while swimming, bullets swishing through the water—is treated with wonder and death-defying humor.) *Saving Private Ryan* returns to the din of battle in a kind of feathered movement—bobbing back and forth, above and below the waves—that returns the film to the louder chaos of the beach. Once back up, as described above, Spielberg uses the resource of Miller's combat deafness to offer two further stretches of respite from the noise.

Once Miller, the central subject for the initial landing sequence, has arrived at the shingle, the other men of the film are introduced by name. Assembled in the safety of the shingle, they are ready and able to, in Miller's words, "get in the war." At the shingle, the sound of the battle changes. The sound of waves, dominant in the first section, falls away. The second, tactic-heavy sequence—thirteen minutes in length—unfolds without the sound of the crashing surf. Miller's group takes out several machine gun positions and, finally, clears a pillbox, opening an exit for vehicles from the beach, completing the initial task of the D-Day landing parties. As a piece of sound art, Spielberg's D-Day moves from wet to dry.

The geography of Omaha Beach—shoreline, shingle, Atlantic Wall— effectively provided a template for Spielberg's sonic mapping of the space, which is never shown in an establishing shot that orients and puts the viewer outside or above the scene. We remain on the ground with the American soldiers or, very briefly, behind the German machine gun. In similar fashion, Kathryn Bigelow's realization of the SEAL raid on Osama bin Laden's compound in Abbottabad, Pakistan, in *Zero Dark Thirty* never completely lays out the space of this similarly iconic battle. Despite claims that the sequence is closely based on actual events, it is—of course—also a highly ordered film experience. Bigelow provides the viewer two modes of sight, distinguished by lighting: natural lighting shots of SEALs moving through the very dark exterior of the compound and the dimly lit interiors of the house and a SEAL POV through green night-vision goggles, a perspective introduced at the close of the insertion sequence, just before the men get out of their helicopters. The SEAL POV is never locked into a specific individual among the group; nor is the POV angle armed (as are several similar POV shots in *Act of Valor*, or as the player in a first-person shooter video game is). Instead, the green-tinted shots secure the viewer to a perspective on the side of the SEALS—we see as they see—and provide an eerie contrast to the very dark norm for the sequence. Toggling between the two lighting schemes makes for variety and close investment with the SEALs.

Bigelow avoids mapping out the compound or explaining the plan of attack. There is no pre-mission briefing—even though the actual SEALs rehearsed on a full-size mock-up of bin Laden's compound—and the layout of the space remains vague, an intentional choice by a director known for making the larger physical context of action sequences legible. Instead, the raid unfolds as a succession of violent acts around the courtyard and in the house, most marked by sonic events breaking the silence. These weapons noises are sudden and sharp and almost all mark progress in the mission. A sequence of explosive breeches of gates and doors are presented in rhythmic terms, with each anticipated detonation counted off by a SEAL. They (and we) can prepare for the explosion. Each breech, with the exception of an unsuccessful one, takes them (and us) farther into the labyrinth, the legendary beast at the center. Short bursts of gunfire also mar the silence of the night, with a clear distinction heard between an AK-47 fired by one of bin Laden's protectors and the SEALs' suppressed fire in the one instance of gunfire exchanged through a closed door. The SEALs' three-fold shots into already-fallen subjects confirm kills in a repeated rhythmic pattern and mark each move deeper into the maze. There's very little chatter on the comms and only one casual exchange among the SEALs. The sequence is all business. Along the way, individual SEALs whisper the names of men assumed to be in the compound: Abrar, Khalid, and finally Osama, that last said three times, as if to conjure the devil. The SEALs whisper a name, then kill its owner. The moment of achievement in the sequence—the shooting of bin Laden—is marked by the repeated utterance of the pro word "Geronimo," embellished with the phrase "for God and country," the only gung-ho touch allowed. The SEALs' reaction to the killing is minimized, too. The shooter says, "I killed the guy on the third floor." Another replies, "Do you realize what you've done?" and they're off to their exfil.

The extremely dry character of the raid in *Zero Dark Thirty* presents Joint Special Operations Command snatch-and-grab warfare in a particular light. The noise discipline Bigelow exercises over the sequence—holding off any musical expression during the raid—does considerable ideological work. The SEALs are consummate and dispassionate professionals. The lack of music or subjective sound plays a part, as does the stripped-down vocabulary of sound effects that moves the viewer through a dark space that's never completely revealed. If we cannot follow the geography of the mission, the SEALs clearly can. Their progress is measured in almost ritualized sounds—prepared breaches, whispered names, triple shots—more than movement through legible space.

Unlike Spielberg's D-Day and Bigelow's Abbottabad, the jungle terrain on which the Vietnam War was fought provides little in the way of legible space. And so, to make extended battles clear to the viewer—and provide shape to the narrative—filmmakers have used contrasting sound areas instead. As with the explosions that move Sledge, Miller, and the mix into and out of subjective sonic areas, Oliver Stone in *Platoon* (1986) uses literally explosive events as punctuation for abrupt shifts to different battle areas and different effects regimes. The final battle offers a good, all-effects example of a combat sequence that is both chaotic and ordered in story terms.

The battle begins with an acousmatic prelude: Captain Harris on the radio with an overrun position as described in the previous chapter. His final "shit" kicks off the combat sequence proper. Abruptly Stone cuts to a Viet Cong unit preparing for battle—one of very few visual representations of the enemy in *Platoon*. Music accompanies this, a haunting, unmetered register using exotic flute and high, resonant wood sticks discussed in chapter 8 under the rubric "veil music." This music cuts out when a trip flare, set some meters outside the perimeter, is set off by the advancing VC. An aesthetic motif of sound and light begins here, with the shushing sound of the flare taking over from the veil-like score cue—music yields to effects on an explosive event.

Chris and Francis see the flare from their foxhole. Rhah crawls in and advises them of "gooks in the fuckin' perimeter." Marking the moment with further sound and light and ratcheting up the tension, a bright white illuminating parachute flare is shot into the sky. It falls slowly to earth, casting a ghostly, strobing light against the green wall of the jungle—one of *Platoon*'s most powerful images—and emitting a remarkable sound, a rapid *boo-boo-boo* that inserts a metaphorical heartbeat charged with fear into the diegetic realm. Vietnam veteran authors have described such flares as making a "squeaking noise" and "swaying and squeaking until it hits the wire, and illumination dissolves."[20] Stone viscerally communicates the fear of imminent attack by way of an authentic battlefield sound. Chris and Francis watch several flares descend (Stone milking the effect). They also hear the sound of enemy chatter, which Chris accurately deciphers as prelude to an RPG attack on their hole. He convinces Francis to get out with him just before said RPG destroys their position. Having survived, Chris charges into the advancing enemy in a noisy firing spree that lasts a long time and breaks the tortured silence of the previous few minutes. Even Francis joins in. Chris marks the end of the firing with the words "fuckin' beautiful," cementing his running attack as a moment of exhilaration. Here, Chris comes close to identification with Barnes and Elias, his two models, both of whom moved freely and without fear toward the enemy. His "fuckin' beautiful" also certifies the buildup and attack as an aesthetic unit in the film that pleases. Stone, here, celebrates the thrill of killing.

Chris and Francis's engagement with the enemy is followed directly by two others of much shorter duration. Bunny and Junior are killed at close range in quick hand-to-hand fights. Barnes charges off; O'Neill hides. These two sequences are separated by the sound-and-light punctuation of an explosion, used as a formal device to change the scene. These explosions are not located precisely on the battlefield. Indeed, the battlefield itself is, by this time, a matter of zones of individual combat rather than a legible space through which the viewer is being moved. There is no larger geometry to this jungle.

A shift in scale follows, offering a broader view of the attack, setting up the air strike that will conclude the sequence in a blast of noise and white light. The command bunker is destroyed by a Vietnamese sapper; Captain Harris, played by Dale Dye, back on the radio, admonishes Lieutenant Wolfe to "hold in place" and fight. Wolfe is killed just after. Harris then calls in the air strike "on my pos" (position) and offers the concluding comment, "Lovely fucking war"—a breach of radio discipline permitted perhaps by the utter collapse of the American position. Stone's script has only the line, "Lovely war." Dye probably added the adjective "fucking"—like his earlier "shit" it's a detail that rings as authentic, delivered with an understatement that keeps it from being an action film tagline. Also holding the one-liner at bay is the fact that the Americans are losing in both instances.

After Harris's final line, Stone cuts to a long shot of the approaching air strike. This special-effects shot has not aged well—the plane moves too slowly—but it offers a break from the relentless battle noises just before the climactic sequence.

Another explosion as punctuation plunges *Platoon* back into close combat. Chris and Barnes come upon each other and Barnes starts to attack Chris. The air strike leaves the outcome of their encounter in doubt and brings the battle to a close with a rising whoosh of sound—as the jet approaches, the pitch of its engines rises, added to multiple explosions, eventually whiting out the image and clearing the soundtrack. Stone described the moment using a rhythmic term: "The PHANTOM FIGHTER JET comes now like a great white whale. One big beautiful monstrous beat of deafening sound."[21] The sound and image of explosions takes the screen to silence and white. The combat sequence ends.

The final battle sequence in *Platoon* can be understood structurally as a kind of musical whole. Sound rises across the sequence, beginning with the battle heard on the radio, then the brief music for the enemy's preparations, then the squeaking flares, followed by the louder hand-to-hand combat—initiated by the RPG strike on Chris and Francis's foxhole, which acts as a starting gun for the shooting and close-quarters fighting. Then the gradually larger explosions of the wider-angle compound sequence, and finally the climax of the napalm strike, which is louder for the long special-effects shot of the approaching plane, which briefly clears the soundtrack of chaotic noise. This ten-minute sequence masterfully paces a terrifying night battle, with targeted lines of dialogue along the way registering both the failures of the battle and Chris's jubilant bloodlust. Music is used only briefly early on: the volume and variety of sound effects instead carries the energy of the relentlessly immersive sequence. We are given little opportunity to pull back.

Many combat sequences do not rely solely on sound effects. Returning to *The Pacific*, shortly after Sledge's episode of subjective sound on the beach at Peleliu, the battle cuts to Leckie, also landing that day. Sledge and Leckie also appear in separate strands of the taking of the Peleliu airfield in episode 7, another frontal assault on a well-defended position. The landings (from Leckie's introduction onward) and the airfield attack—respectively thirteen and eleven minutes in length—both enhance continuous noisy effects with a layer of sound that sits between effects and music. Of indeterminate origin, not located in the diegesis but not easily traced to conventional musical instruments, these dissonant, metallic, harsh, biting, swelling, nails-on-a-chalkboard eruptions of sound thicken the effects-dominated combat soundscape. These pulses of sound are neither melodic nor metrical (in a rhythmic pattern), but textural. In context, this sonic layer suggests several metaphorical interpretations: fear or terror at possible sudden death; the predicament of individual soldiers caught in the realities of what Robert Leckie in his 1957 book *Helmet for My Pillow* called "mechanical war"; psychological resistance to going forward; a more general expression of the danger of battle; the momentousness of the combat experience itself, a kind of generalized and sustained "yes we're being shot at."[22] (These sounds are notably absent from Basilone's battles.) *The Pacific*'s metallic swells provide an affect-heavy enhancement to combat scenes where the objective is rather simple: survive a run across open ground. They effectively capture the existential peril of "mechanical war," but also remind the viewer that they are watching a representation, providing sonic leverage allowing for reflection, intruding on an image track that could have easily been mixed with only diegetic sound. (Two earlier PCFs deploy such sounds deep in the mix during battle sequences: *The Thin Red Line* and *Flags of Our Fathers*.)

Metallic swells sit just this side of scoring. In a few cases, effects-centered combat set pieces bring in musical cues. As noted, this happens briefly at Basilone's death on Iwo Jima. Contrasting examples from *Casualties of War* show how different sorts of music can direct viewer attention in the midst of combat action. In the opening scene, Eriksson is blown into a hole, which opens into a tunnel system beneath his feet. Meserve goes back to find and save him. Music comes in when a Vietnamese soldier in the tunnel begins crawling toward Eriksson's legs. Meserve simultaneously advances on Eriksson aboveground. The sequence—with bad guy and hero approaching a trapped "damsel in distress" from different directions—reads as drawn from a horror film. Composer Ennio Morricone adds pulsing, heroic action music, heightening the race to Eriksson with rising, tense figures. Meserve even gets a few fanfares. Pulling Eriksson free at the last second, the two Americans come under fire from an enemy soldier perched in a nearby tree—another instance of spatial variety in the battle. (Meserve, firing on automatic, shoots him down, crying, "Get some, motherfucker"—thereby introducing his signature line. Eriksson cries out, not all that convincingly, "Yeah, you fuck." His general inability to talk tough is introduced here.) The sound of Meserve's automatic fire drowns out the music. Once again, the start of weapons fire brings a musical cue to an end—a standard sonic punctuation technique across the PCF and action genre more generally, suggesting some of the conventions of spotting musical cues within a noisy effects-driven context.

Music in *Casualties of War*'s opening battle sequence works to melodramatically underscore Eriksson's vulnerable position and to misdirect the viewer on Meserve's character. Meserve will not be the hero in this film. Indeed, the music heard in the opening battle is never heard again. Rather than dwell on the American soldiers, Morricone devotes the expressive power of his score to the character of Oanh. By the film's second battle scene, the ground has been prepared for Morricone to interrupt combat action with music that directs our attention to Oanh. Having been raped then stabbed repeatedly and left to die, Oanh rises and begins to walk through the firefight, still trying to escape her American tormentors. As she rises, in a context of noisy firing, a lone flute enters and overwhelms the other elements in the mix. Morricone's music intrudes on the battle as a kind of moral pointer: Oanh is what matters here, the score says, not the men shooting at each other. Indeed, after Oanh is shot by the Americans and falls to her death, a sequence of explosions follows, destroying both Vietnamese on the river's edge below and Americans approaching in a patrol boat. These explosions, typical conclusions to a combat sequence, are narratively meaningless and suggest nothing but waste.

Two sonic options for representing protracted, methodical combat—one using music, one only effects—are used during the long, deadening process of clearing Japanese troops from the mountains and caves of Peleliu and Okinawa in *The Pacific* episodes 7 and 9. Episode 7 opens with a musicalized montage of the two-month process of clearing the caves of Peleliu. A second such sequence occurs about halfway through the episode. Various images of combat cross-fade with shots of Sledge's exhausted face. This impressionistic representation of the long battle is shaped to match two complete iterations of one of the series' main musical themes. The viewer is enjoined to think about the battle in general terms to the sound of pathos-heavy music. Duration is again suggested—there's no direction here, no sense that the battle is won or even progressing—but not as a matter of enduring constant terror (as with Leckie's night battle). Instead, the theme for Sledge on Peleliu is one of fatigue and soul-numbing violence. As his soul grows hard, the soundtrack grows sensitive—a cue for reflection, rather than identification, on the viewer's part. We are not called upon to walk with Sledge or concern ourselves with the tactics and dangers of cave clearing. We are to reflect on the fact of these actions rather than the actions themselves. As the images get ugly, the score grows beautiful. The music, of course, offers Sledge himself no comfort.

With Basilone dead and Leckie out of the war, episode 9 remains entirely with Sledge as he and his group, including some replacements, endure the last days of the Pacific War. It is the bleakest episode in the series, forcing the viewer to gut it out for one last hour before the sentimental final episode, set entirely back home in the United States. The Japanese continue to fight on Okinawa but added to the mix are civilians caught in the cross fire—including one used by the Japanese as a suicide bomb. The episode cuts relentlessly and jarringly from day to night, from hot sun to pouring rain, from fierce battle to exhausted rest. Lyrical music—reacting too freely in earlier episodes, where it responded at times to any suggestion of emotion—falls silent. The only slightly musical sound for the first fifty minutes of episode 9 is the return of the ominous metallic hums and surges heard during the airfield sequence. It's a merciless stretch of combat filmmaking—image track and sound track alike cut with a dull knife. The episode climaxes when Sledge finds himself face to face and alone with a severely injured Okinawan woman, who asks him to shoot her and put her out of her misery. (Parallels with the end of *Full Metal Jacket* [1987] are ripe.) He cannot do it and instead takes the dying old woman in his arms, holding her until she passes away. Only here does the comfort of string music return to the score, in a cue titled "Sledge's Humanity."

"It was like having to watch a movie: the movie was inside his head, and he couldn't stop it by closing his eyes."[23] So Roxana Robinson describes the experience of post-traumatic stress disorder (PTSD) in her 2013 novel *Sparta*. Of course, one reason you can't stop a movie by closing your eyes is because your ears remain always open. The sounds of battle intruding on a soldier's life after combat prove a potent and flexible trope in the PCF. Sound effects have expressive, character-building power in these cases. In *We Were Soldiers* (2002), Moore anticipates the sounds of battle while still stateside, looking at books about the French defeat in Vietnam (where he's shortly headed) and Custer's Last Stand (his new airmobile unit having been dubbed the Seventh Cavalry). Battle sounds in Moore's imagination hint at the futility of the American war in Vietnam before it has even begun. Joe Enders in *Windtalkers* (2002) hears again and again the cries of the men he led into death on a mission only he survived. Their melodramatic cries—"God damn you, Joe Enders"—undercut the moment, marking a lapse in tone. Memories of battle mistakes rendered as sonic flashbacks recur across *Born on the Fourth of July* (1989). One such moment occurs in public. While addressing a July Fourth crowd at a hometown pro-war rally, Kovic hears a baby's cry. The small sound stands in for an earlier noisy combat episode, when Kovic and his squad accidentally killed women and children. The sound of an unseen helo, probably only in Kovic's head, adds another weighty element to the soundtrack at this uncomfortable juncture. For these moments, the movie audience has access to Kovic's subjective experience even as the crowd at the rally cannot know why the returned war hero breaks off his speech. The humiliation of PTSD, a condition often invisible to others, comes across in a context generating great sympathy for Kovic.

Stacey Peebles identifies a continuing motif in narratives of veterans from the Gulf and Iraq Wars. "In these stories, many veterans return to the United States to discover the unexpected pain of being 'in between' war and home, not able to fully exist in either state."[24] In the PCF, as exemplified by the above scene from *Born on the Fourth of July*, disjunction between sound and image often marks this "in between" area, with the image track at home and the soundtrack at war. *American Sniper* offers very few clues to Kyle's inner life. He doesn't talk about the war much beyond uttering certitudes that those around him find difficult to engage with. But Kyle's experience of sound on his rotations home does offer evidence for unseen wounds. Twice he reacts with alarm to explosive noises of daily life—a lawnmower, a pneumatic drill—as if they were the sounds of war. These moments enact veteran poet Brian Turner's lines "Cash registers open and slide shut / with a sound of machine guns being charged."[25] But the most developed expression of Kyle's post-combat state of mind comes in a three-shot sequence when the warrior appears almost completely lost to himself.

It begins with Kyle in his living room apparently watching television: initially framed with Kyle facing the set, its screen unseen. The sounds of battle, likely a war film, are heard. Eastwood tracks smoothly to a different angle where the television screen is visible. The set is off. The sounds are in Kyle's head. Taya interrupts and on a hard cut to her the war sounds cut out. She exits and, on a return to Kyle watching the black screen, the sounds of war cut back in. Here, in a rare moment, *American Sniper* suggests by way of sound effects the lingering effects of combat on a warrior otherwise presented as almost completely locked down.

7. Helicopter Music

> The percussive sound of the rotors is so powerful that it enters far into the realm of the haptic: as it approaches, it is deeply felt as well as heard. . . . The low percussive roll of the rotors can force one's chest to vibrate in concert with its pulses.
>
> —J. Martin Daughtry on the sound of a Chinook helicopter (2015)

> If you closed your eyes you seemed to hear music, military but atonal like tinnitus.
>
> —Novelist Martin Amis riding a Cobra helicopter (2007)

> As night fell, the operations center hummed with serious, focused activity. Soon, the rumble of helicopters and aircraft, some throaty, some a high whine, bounced across the darkened gravel and off the cement walls and barriers of our compound. The sound grew in layers, building like a chorus singing a round, as one set of rotors, propellers, or jet engines came alive, joined the cacophony, and then departed the airfield. Gradually, the chorus dissipated until silence returned to the darkened base.
>
> —GENERAL STANLEY MCCHRYSTAL, *My Share of the Task* (2013)

If films about the Vietnam War posed a narrative challenge—war stories ending in defeat—they offered in consolation an instrument of war with all the attributes of a movie star: the helicopter. These undeniably attractive, awesomely powerful machines defy gravity, come to the rescue, and fall tragically to earth, all while moving at a speed that welcomes capture by the camera's eye and in a shape that fits the wide-screen frame. The dynamic physicality of the helicopter proves a natural for the cinema. The problem is the helicopter's voice: it's too loud.

We Were Soldiers (2002) tells the story of the early days of airmobile combat and the first major battle in Vietnam that tested this new strategy for moving men on and off the battlefield. Early in the film, Lieutenant Colonel Moore, played by Mel Gibson, addresses the group of young officers he will train in the innovative and untested techniques of airmobile combat. The setting is a hangar on a military base: a large, resonant, silent space where Moore's every footstep sounds out crisp and clean as he strides toward his men standing at attention. We are not, of course, actually hearing the sound of Moore's (or Gibson's) footsteps. A Foley artist—a maker of performed sound effects—created those sounds in postproduction, shaping their volume, tone, and timbre in collaboration with other effects designers and mixers as if they were musical notes: as, indeed, they arguably are in the overall sonic structure of the scene. The steps act as a prelude to a short speech which, in turn, sets up the climax of this little piece of sonic fantasy. Moore says, "At ease, gentlemen. Welcome to the new cavalry. We will ride into battle. And this will be our horse." He points to the open hangar doors and a Huey helicopter makes a sudden, whooshing entrance, pauses to acknowledge Moore's salute, then flies off. Two further Hueys fly past the hangar door in close formation.

Assuming a Huey designed to carry groups of men into battle and to engage the enemy with its own rockets and guns makes *at least* as much noise as the average two-seat traffic copter or a slightly larger air ambulance—two kinds of helicopters civilians have occasion to encounter in daily life—the suddenness with which this trio of military choppers appears suggests, contrary to lived experience, that a helicopter can sneak up on you. In Hollywood, helicopters can turn sharp corners without announcing their presence sonically; they can take us by surprise. Defiance of the audience's lived acoustic experience with the helicopter has proven essential to making a film with helicopters in it. Indeed, for most war movie audiences the closest they will ever get to a military helo is at the movies. Hence the cinematic representation of the helicopter becomes, for most, the reality of this essential tool in the American arsenal from Vietnam to the present.

This chapter considers strategies used in the PCF to modulate and musicalize the sonic representation of this irresistibly audiovisual totem of modern combat. This mechanical movie star has played too many parts and enacted too many scenarios of success and failure for a single theory to explain its use in the PCF. It is worth noting, however, that since the introduction of the helicopter in the early 1960s, the US military has failed to score a decisive victory in any major conflict relying on the helicopter. The ultimate utility of the helicopter to win wars remains in question. The *We Were Soldiers* book and film, unsurprisingly, do not question the efficacy of the helo—nor does any PCF. Hollywood audiences have tacitly embraced the ambiguous nature of the helicopter: we love to see them fly, and we love, in equal measure, to watch them crash. The helo in the PCF has developed along an odd line of fetishization of a technological marvel that can fail so easily and, in audiovisual terms, so spectacularly. This chapter considers the sonic representation of the helicopter in PCFs set in Vietnam and the Greater Middle East. I pay special attention to scenes of soldiers inside helicopters and to how music has been used to shape the cinematic experience of helicopter-borne battle. Film form often follows musical form when helos take to the skies on-screen.

MODULATING OR NOT

Helicopters are deafeningly loud at close range and announce their presence sonically at a great distance. For example, military helicopters taking the president from the White House to Andrews Air Force Base regularly trouble the quiet at the Vietnam Veterans Memorial. Have their rotor sounds ever activated a veteran's sense of danger? Even nonveteran visitors to the Wall might register the association between the war in Vietnam and the sound of a helicopter, a connection forged, in part, in the cinema. Helo sounds are part of the Wall's soundscape. But the sound of a helicopter at any given moment on a film soundtrack is completely at the discretion of the sound department, mixer, and director. In the movies, rotor sounds can be turned off or turned down—or in sound mixing terms, modulated.

Some films do not modulate helo sounds and instead use the volume and changing pitch of rotor blades for storytelling purposes. *84 Charlie MoPic* (1989) does this as a matter of course: the film's rigorously diegetic, point-of-audition soundtrack forces the mixer's hand. In the process, the viewer learns something concrete about the soundscape of the combat zone for American soldiers. As one soldier says, "One good thing about the Nam. If you hear a chopper, you know it's ours." In the rescue sequence in *The Deer Hunter* (1978), three Americans fleeing downriver from a prisoner of war camp hear a chopper in the distance. Its volume grows as the chopper approaches, is deafeningly loud during the attempt to pick up the three, and leaves a lonely silence in its wake when it departs with two of the three Americans not rescued. The soundtrack remains locked into the audio perspective of the three men: the communications net is never brought into the mix; shouts in the helo's bay are not audible. It's as close to live sound as possible. Director Michael Cimino said of *The Deer Hunter*, "Look, the film is not realistic. It's surrealistic. . . . Literal accuracy was never intended."[1] And yet in this scene, Cimino presents the helicopter accurately in terms of aural perspective and to great dramatic effect. The difficulty of effecting a rescue mid-river in hostile territory is brought home to the viewer in a sustained, excruciating sequence that unfolds in something close to real time, judging from the way the helo rotors mark out space. The combat journalist Peter Arnett had harsh words for *The Deer Hunter* but praised this scene: "There are a few real moments. Helicopters bucked and strained against the pull of gravity."[2] The mix aids the viewer in understanding the challenges inherent in the machine itself.

At close range, helicopters make a terrific amount of noise and wind. Firsthand witnesses regularly compare helos to the most violent forces of nature. Vietnam memoirist Philip Caputo describes "each chopper creating a miniature hurricane. . . . The noise was terrific. . . . We had to shout to make ourselves heard."[3] The military historian Samuel Zaffiri writes, "The men nearest the ships had to hold onto their helmets to keep from having them knocked off by the fierce pull of the ships' rotors and to cover their faces with towels to keep from choking on the dust swirling through the air. . . . Eventually the entire PZ was covered with lift-ships and a tornado of sound."[4] Iraq veteran Colby Buzzell notes on riding a Chinook, "It was extremely loud inside and I'd lost my earplugs, so I got a couple expired cigarette butts out of my pants pockets and shoved them in my ears."[5] Veteran author William Pelfrey describes the experience of "insertion into unknown terrain by helicopter" in Vietnam: "It meant flying in the door-less cargo bay with your knees jammed into your chest or your legs dangling above the skid, unable to hear anything going on between the pilot and Higher or anything happening on the ground."[6]

Were cinematic helos to sustain a strictly realistic mix, the helo's role as a setting would be limited. Modulating rotor sound down proves essential if dialogue scenes are to be played inside a helicopter. In *Full Metal Jacket* (1987) an entire conversation occurs across the open bay of a helicopter in flight. (In preproduction meetings Stanley Kubrick insisted that the film—shot in an unconvincingly faux Vietnam re-created in Britain—"must" include at least one helicopter.)[7] Joker and Rafterman are being taken to the battle zone. A door gunner fires on automatic at fleeing Vietnamese below, yelling repeatedly, "Get some." Joker and the gunner engage in a brief conversation that touches on body counts and the killing of women and children. Neither man leans toward the other while speaking—there's no sense of strain over the rotor noise—and Kubrick never establishes their physical proximity within the chopper, likely because there would be no camera position for a shot showing both. In context, the conversation works within the stylized nature of Kubrick's film, which presents aspects of the war rather than representing the war under any cinematic notion of realism. In practice, creating the soundtrack for the scene proved difficult. Postproduction notes register concern about the sync between the door gunner's words and lip movements, revealing that the scene was shot without "clapperboards or handclaps"—in other words, without live sound—and the dialogue "all lip-synched" in post.[8] The mix in this scene is an entirely synthetic confection. In *Apocalypse Now* (1979), as Kilgore's flotilla of choppers rides into battle, characters in a similar context catch each other's words without seeing each other's mouths, a choice of staging for the camera that makes Kubrick's face-to-face exchange almost credible. In *Casualties of War* (1989), Meserve gives his mortally wounded friend Brown advice just before he is medevaced: "I'm gonna hypnotize you," he says. Meserve speaks at close range to Brown, who lies on a stretcher in the open bay of a helicopter. In the real world, Meserve's words would be incomprehensible. Indeed, in these shots the director Brian De Palma does not show the chopper's rotor blades, and the lack of rotor wash suggests that the scene was shot without the helicopter turned on (even though the helo lifts off seconds later on a cut to a wider angle). Inevitably this influences how Sean Penn as Meserve delivers his lines. He doesn't shout nearly to the extent he would were the blades turning. And if Penn's lines were looped, he was not coached to yell. Modulating rotor volume downward for all these dialogue scenes falls within the so-called intelligibility imperative, a maxim of Hollywood sound mixing dictating that dialogue trumps all other sounds. Such scenes could, perhaps, be blamed on screenwriters who insufficiently imagined the environment in and around a helicopter. Indeed, two Vietnam films written by veterans of the war, *Platoon* (1986) and *Hamburger Hill* (1987), do not indulge in such sonic conceits.

But dialogue shouted over chopper blades can also serve important expressive purposes. Returning to Brown's death in *Casualties of War*, the soundtrack mix in this sequence decisively favors Ennio Morricone's score. The last we see of Brown, he's alone in the chopper in flight over an expanse of green jungle, shouting over and over the words, "I'm a armor-plated motherfucker." The helicopter, its rotors modulated down, becomes a stage for the young soldier's final, defiant speech act, which is haloed by elegiac music that signals Brown's imminent death. Of course no one in the film is there to hear Brown's cries; only the audience for the film, whose needs are served by the mix.

Conversations in helicopters in films set from the 1990s forward generally get around the sonic absurdity of modulating the rotors by putting everyone in the chopper—and, of course, the film audience—on the net.

HELOS AND STORYTELLING

As vehicles of storytelling, helicopters have been used in targeted moments to articulate larger themes or mark important transitions in a film narrative. These examples show the flexibility of the helo as a visually and sonically dynamic narrative element, often deployed in a quasi-musical manner.

Early in *Three Kings* (1999), an officer orders Archie, played by George Clooney, to get his act together. The Gulf War has supposedly been won, although Archie resists the official story, saying, "I don't even know what we did here." The officer answers, "What do you want to do, occupy Iraq and do Vietnam all over?"—a line that precisely dates *Three Kings* as pre-9/11 and carries retroactive irony for twenty-first-century viewers. The rotor wash of the departing officer's helo literally leaves Archie in the dust. Here, the helicopter—its roar and downwash enhanced—innovatively expresses the power of the military to suppress individual thought, setting up the plot trajectory where Archie elects to defy US policy and help a group of refugees escape Saddam Hussein's Iraq.

A similar moment—a helo rising over a seated soldier—carries very different meaning in *Hamburger Hill*. Late in the film, the strong yet sensitive black medic Doc has been wounded and is awaiting medevac. Frantz and Motown chat with Doc, who urges them to "take the hill." Whether Doc lives or dies remains ambiguous. He exits the film still seated. A helo very nearby begins to ascend and Frantz and Motown embrace Doc to protect him from the downwash. They form an interracial sculptural group—echoing the statue "Three Soldiers," a more traditional bronze added to Lin's austere Wall—which comes across, as one critic noted, with "the chattering requiem of a helicopter . . . [as] one of the most powerful scenes in the film."[9] In some respects this moment simply transfers the power of the helo to the power of the narrative juncture (Doc's ambiguous departure from the film), outside any easily definable meaning. The relatively open nature of certain classes of sound—whether instrumental music (lacking the interpretive pointer of a sung text) or the "chatter" of a helo—marks an occasion when meaning is evident but, perhaps, not all that easy to pin down. In such moments lies the power of both movie music and the music of the helicopter, a unique instrument of war. To recall Walter Murch's words from chapter 2, "Sometimes sound can become almost pure music; it doesn't declare itself as music which is actually one of its advantages because it can have a musical effect on you without you realizing it."[10]

In *Green Zone* (2010), Captain Roy Miller and his team hunt down supposed caches of WMDs (weapons of mass destruction) in Humvees, spending almost the entire film outside the Green Zone: speeding down Route Irish, the dangerous stretch of highway leading to the airport, or trapped in traffic on the streets of Baghdad. The Humvee as death trap is a frequent motif in Iraq War films but as a vehicle it, like the jeep in World War II films, has little cinematic personality. Special Forces guys, working at loggerheads with Miller's allies in the CIA, interrupt Miller's efforts from their superior vehicles—three gorgeously lethal Hueys. The Hueys initially appear in a brilliant shot that has the lead helo enter from the bottom of the frame, suddenly rising above a small hill, its sound taking Miller and the viewer by surprise in another instance of Hollywood helos turning a sharp corner. These helos are agents of military and administration forces trying to hunt down a former Iraqi general (who, the story goes, told the Americans before the invasion that there weren't any WMDs). Their extravagant helicopter-borne entrance and exit, interrupting Miller's progress on his own mission, casts his search for the truth as an underdog's quest against a not exactly evil but definitely morally bankrupt counterforce supposedly on Miller's side. (Tensions between regular Army, Joint Special Operations Command, and the CIA are part of *Green Zone*'s effort to show the complications of the GWOT.) Late in the film, Miller is taken hostage by the general. All American forces in the area, led by the Special Forces guys in their helos, who had been tracking Miller, descend on a Baghdad neighborhood to rescue him. Here, the JSOC helos are repositioned narratively as coming to Miller's rescue and—in a nod to the inevitable ambiguity of these terrifying but vulnerable mechanical dragons— one of the choppers is shot down, providing a fiery explosion in the midst of a long foot-bound chase sequence through dark streets and alleys. This final sequence, shot months after completion of principal photography, replaced an earlier version. In their quest to bring *Green Zone* to an appropriately dramatic action conclusion, the film's makers settled on a helo crash as a compelling narrative element.

Acousmatic (heard but not seen) helos are used as a kind of dramatic scoring in several films, sounding in the distance at tense moments of otherwise quiet dialogue or action, their sound shaped for dramatic effect. Such moments deploy a kind of effects-made music that, crucially, does not require the use of sounds the audience interprets literally *as* musical sound—in other words, this is *musique concrète* (electroacoustic music made from everyday sounds) that is dramatically targeted for specific narrative ends, just as movie music typically is. In *The Deer Hunter*, the transition from the United States to Vietnam is sonic. The men end the film's long opening sequence in Clairton back at John's bar. Not saying a word, John sits at the piano and plays Frédéric Chopin's *Nocturne in G Minor*, op. 15, no. 3. All listen in silence. A lengthy shot slowly moves from face to face. We listen with the men while also watching the men listen. In the words of the script, "It is so unexpected, so beautiful, so sensitive, that it startles and moves us." As the nocturne ends, helo sounds enter the soundtrack, growing and growing. The image track cuts abruptly to helos destroying a Vietnamese village and the three main characters in combat. The helo sounds are revealed to be prelapped. Cimino originally scripted a jarring sound match between a train outside the bar and jungle noises and Hueys in Vietnam.[11] In the finished film, he opted for a gentler transition, ending the Clairton section on a quiet note, the helo rotors working like disturbing music disrupting the silence after the "unexpected" Chopin.

Often acousmatic helos never make an appearance in the image track. Such cases assume a battle zone where helos are common, a soundscape open to the expressive option of subtly "scoring" tense moments with effects-made "music." *Hamburger Hill* uses unseen helo sounds as a recurring theme. Early in the film, Sergeant Frantz makes an illustrated speech to some cherries about the nature of the enemy. A Vietnamese man on the American side plays the part of "Han the sapper," demonstrating how the enemy can move silently past barbed-wire barriers. Cutting between images of Frantz talking intently and "Han" sliding through the mud, the soundtrack features Frantz's forceful yet quiet voice given added weight by the sound of an unseen helicopter, which can also be heard as the nervous heartbeats of the listening cherries. Similar unseen helos add tension to an exchange of looks between Doc and Frantz after the death of a black soldier assigned to walk "point" and when Frantz angrily confronts a reporter late in the film—in the last, the helo sounds rise in volume across his short speech. Dispensing almost entirely with musical scoring, *Hamburger Hill* instead uses helo noise to add weight to tense moments.

The Kingdom (2007), *The Hurt Locker* (2008), and *American Sniper* (2014) all also deploy unseen helo noises to ratchet up tension. In *The Kingdom*, helo noise attends an untrustworthy State Department official who tries to cut short the FBI team's quest for the truth. In *The Hurt Locker*, unseen helos fly overhead during James's awkward call home—he never says a word—and when he sneaks into an Iraqi vendor's truck and demands that the man drive him into town. In *American Sniper*, just as Kyle learns that his brother has been deployed to Iraq, a helo flies over, thickening the soundtrack and hinting at emotion in this otherwise stoic character. These films have highly contrasting scores in terms of style and extent, yet all three, at similarly small dramatic junctures, opt for unseen helo noise as a way to subtly turn up the tension.

Along a more specific expressive line, Oliver Stone exploits an unseen helo as an analogue for the human heart in *Platoon*. Early in the film, the cherry Gardner dies of a sucking chest wound during a night patrol. The stilling of Gardner's heart is expressed sonically by the slowing sound of a helicopter's rotors. From a literal perspective, this chopper is flying away from the scene—a strange choice given the medic's shouted assurance to Gardner that a "bird's on the way." But the sound—convincingly diegetic given all the helos flying around these films—works metaphorically as the ebbing of Gardner's life. He dies on the soundtrack.

In all these instances, the distinctive and flexible sound of the helicopter serves as effects-derived scoring, as helicopter music made entirely of the sound of the helicopter and used diegetically within combat contexts to mark important moments, in the manner of movie music but without recourse to identifiably musical sound.

MUSICAL INSERTIONS

The most extreme modulation of helicopter noise has been a musicalizing of the machine: supplementing or replacing the helo's characteristic sound with music. Musical sequences set the helicopter dancing and, with significant possibilities for narrative filmmaking, musical content displacing helo sounds has been used to fill out the inner lives of the soldiers riding into or out of battle. Below I consider parallel helicopter sequences in four films: *Hamburger Hill, We Were Soldiers, Black Hawk Down* (2001), and *Zero Dark Thirty* (2012). Each uses an extended helicopter ride to mark a large-scale narrative inflection point: the move from preparation to battle. Each uses music to position the viewer relative to specific soldiers and wars. This comparative analysis adds specific musical content to earlier discussions of the changing figure of the American fighting man across the four PCF cycles.

The transition from base camp to jungle warfare in *Hamburger Hill* is marked by an extended sequence that forms an ABA structure. A group of helicopters arrives to pick up the men, to the overwhelming sound of rotors. Director John Irvin's low-angle shots emphasize the force of the helos' downwash, as if the camera, like the men, is ducking as low to the ground as possible. The flight into the jungle features an almost complete playing of the Animals' 1965 hit "We Gotta Get Out of This Place." This combat insertion is a song scene. Critics found the choice of music either an "exhilarating one," a rare moment when "Irvin allows himself to poeticize the action," or "so ham-fisted [the idea] could have been hatched in the Schwarzenegger brain."[12] Such polarized reactions could just as easily be activated by the song itself, carrying a different resonance for different viewers, as by its use in the film. As Bruce Springsteen said of the Animals, mentioning this track specifically: "For some, they were just another one of the really good beat groups that came out of the sixties. But to me, the Animals were a revelation. The first records with full-blown class consciousness that I had ever heard."[13] Kevin Hillstrom and Laurie Collier Hillstrom's *The Vietnam Experience* (1998), an encyclopedia of literature, songs, and films, describes "We Gotta Get Out of This Place" as "undoubt-edly" among the most popular rock records for soldiers in country, "an anthem of sorts among American troops. . . . Years after its release, the song remained a staple in Saigon jukeboxes and fire base tape players."[14] As noted in chapter 2, Irvin selected this track because he felt Francis Ford Coppola's use of "Ride of the Valkyries" in *Apocalypse Now* grossly mis-represented the demeanor of soldiers he observed in Vietnam heading into the battle on helicopters. Irvin remembered the Animals' song as "an anthem" for the young soldiers he met as a combat documentarian in Vietnam: "I used it when they were going into battle to suggest not just a sense of anxiety and foreboding, but also as a riposte to Coppola putting Wagner into his film which I thought was singularly inappropriate."[15] When the choppers descend into the jungle and the men disembark, the music fades out and rotor sounds fill the soundtrack once again. The depar-ture of the choppers leaves the men in the silence of the jungle, a silence enhanced by the preceding noise of the helos—an effect repeated in several Vietnam films and at the insertion of the four SEALs in *Lone Survivor* (2013). The insertion sequence in *Hamburger Hill* is a highly structured ABA: rotors, song, rotors; effects, music, effects. After passing through this portal, the film remains in the combat zone to its conclusion.

"We Gotta Get Out of This Place" begins with a bass riff that effectively announces the song for some audience members. This figure sneaks in on the soundtrack only after the scale of the shot puts the flotilla of helicopters at a distance where the lowered volume of their rotors makes auditory sense. We are not asked—at this point—to believe that the Animals can drown out seven choppers. The magnificent long shot of the choppers that follows—cut exactly to the track's intro and the first phrase of the verse—was possible only because *Hamburger Hill*, a modestly budgeted film, sought and received Pentagon support. (Similarly grand reenactments of the helicopter-heavy American war machine were simply outside the range of possibilities in *Platoon* or *Full Metal Jacket*, neither of which had Defense Department assistance, and both of which made do with just a few rented helos.) On the second phrase of the verse, however, we are put abruptly onboard the chop-pers, in close proximity to the men—but we cannot hear the rotors. The song's verse accompanies intimate shots of the men, whom the viewer knows as individuals by this point, having endured thirty minutes of train-ing with them in the film's first part. The combat genre, with its target audi-ence of men, is visually obsessed with faces: men, in short, are invited to look closely at other men, to read their faces, to intuit their feelings. Music mod-erates some of the embarrassment so intimate a sharing of male feeling might engender. For these shots, rotor sounds are completely absent. We attend to the men in the helicopters in a state of diegetic silence: music floods the soundtrack. At the start of the song's chorus, the film cuts to a grand shot of helicopters in the jungle, a beautiful image of the helicopter as pure cinema, smoke curling around the rotors, revealing how these war machines create powerful circular currents in the air. These shots get a touch of rotor sound, as if the sonic suggestiveness of visible rotors demands some accounting for on the soundtrack. On cuts back to the men, the sound of the rotors drops out abruptly. This is a highly stylized soundscape in a film otherwise committed to down-in-the-mud realism. The structure of this sequence can be understood by way of musical form: the individual men "sing" the verse as a series of solos, then join together for the chorus, whose lyric takes on ironic meaning as they are heading into battle rather than "outta this place."

Without suggesting outright fear, Irvin's song scene insertion generates tremendous sympathetic energy toward the men of *Hamburger Hill*. They are strong—and so is the music—but they are without recourse, committed to the sphere of battle regardless of the meaning or meaninglessness of the war. Rock as a music of rebellion works ironically here—if gently so.

A similar although briefer musicalized helicopter-borne insertion in *Black Hawk Down* also marks the move from base camp to battlefield, initiating the action portion of the film. The sequence, and the mission, begins with repetitions of the code word "Irene" and shots of helos being readied for takeoff and men loading, images that wouldn't be out of place in a recruiting commercial. Over the sound of the rotors, the guitar solo opening Stevie Ray Vaughan's cover of Jimi Hendrix's "Voodoo Child (Slight Return)" enters as scoring. Before liftoff, a short, completely unrealistic dialogue scene occurs between General Garrison and some of the Rangers at the open door of a Black Hawk. Garrison, as played by Sam Shepard, stands outside the helo to offer parting words. He does not shout or even speak loudly. He does, however, plant the film's tagline, making his final order of the day the words "no one gets left behind." Soldiers Grimes and Eversmann seem much more aware of the context: actors Ewan McGregor and Josh Hartnett visibly shout. (Almost certainly all this dialogue was added in postproduction, where the balance could be adjusted.)

To the sound of Vaughan mixed strongly against helos and Humvees, the convoy and choppers set off. (An officer closes out the dialogue with a businesslike "Let's roll," a line that would doubtless have carried more resonance had the film been made after 9/11.) The Humvees leave first. A rising crane shot has individual vehicles enter from the bottom of the frame, several roughly in time to the beat set down by Vaughan's track: the sound of each Humvee is folded into the rhythm of the music. The synchronization of military vehicles and music is locked into place here. Next, the "armada"—to use the script's characterization—of Black Hawks and Little Birds lifts off with the sexy precision of an air show. (The moment recalls Michael Herr's experience of helos in Vietnam: "You had to stop once in a while and admire the machinery," he writes, and quotes a captain who says of helos hovering "like wasps outside a nest": "That's sex. . . . That's pure sex.")[16] The accumulated roar is presented as enough to vibrate a Somali warlord's glass of tea nearly off the table in a nearby hangar.

"Voodoo Child" fills the soundtrack during the next segment, standing in almost entirely for the actual sound of the armada. A Somali boy hears and sees the helos approaching and calls in to the city to warn the militias that the Americans are coming. He doesn't need to speak, but simply raises his cell phone to the sky as the helos pass over. On a hard cut to the other end of the call, the Vaughan track cuts out and a new, jangly beat—called "Tribal War" on the score album, associated with the big black militia leader who will shoot down the Black Hawks—drops in abruptly. This music plays behind Somalis taking up their weapons and continues on a cut back to the American Humvees and helos heading into battle. Such sudden shifts in the rhythmic pace and texture of the score are central to the film (as discussed in chapter 9). What's noteworthy here is that both musics—the Vaughan rock track and composer Hans Zimmer's "Tribal War"—play behind images of both the Americans and the Somalians. While the latter music in particular is closely associated with the menacing Somali fighter who shoots down American helos, the goal here seems to be to set up the story—with parallel preparations for battle on both sides—and to keep the musical conceit going, to stitch together the crosscuts between Americans and Somalis with an action-anticipating beat.

But before the battle begins, director Ridley Scott offers an awesomely beautiful spectacle of helos in flight. These breathtaking shots of multiple helos piloted by skilled military pilots high above a pristine African beach would have been impossible to make just months later. (Even in a pre-9/11 context, the Pentagon provided so many helos and pilots to the location shoot that joint airborne exercises with the Moroccan military were scheduled to assure the government of Morocco that this was just a movie and not an invading force.)[17] In the air with the armada, all diegetic sound falls away, yielding to a shimmering musical rendering of the rotors by Zimmer as the helos fly in formation above the waves. The moment is graceful, free, and exhilarating, evocative of a luxury car commercial—and an outright fantasy in sonic terms. In reality, these machines make huge amounts of noise, a point made moments earlier by the warlord's glass of tea and the boy's cell phone. But for this suspended moment, just before troops are committed to the streets of the city and the downhill slide of the mission begins (which it does almost immediately), *Black Hawk Down*'s American audience gets to imagine the helicopter as a machine that makes music but not noise, as an instrument of power that puts our soldiers above the mess of a foreign nation. Of course this conceit is a lie: as the title of the film admits, some of these Black Hawks will fall to earth. Still, the sexy presentation of the choppers heading into battle can't help but celebrate their supposed power.

Zimmer's score also touches emotional points in an exchange of looks between Eversmann (in a Black Hawk) and Hoot (on one of the benches outside the cockpit of a Little Bird). As noted earlier, the pair share three talks, distributed across the film, during which Hoot offers no-nonsense advice to the still somewhat soft Eversmann. On the insertion flight, the pair share a sort of mind-meld moment. Zimmer compacts the substance of their talks into a sizzling stinger that lends Hoot's hard stare at Eversmann a forceful edge, much more effective in context than any cry of "get some." *Black Hawk Down* allows its warriors to show some apprehension while headed into battle. Only after 9/11 does Hoot's fearless demeanor become the default posture of elite soldiers, making spine-stiffening stingers unnecessary.

The insertion sequence in *We Were Soldiers* uses a combination of pre-existing music (rerecorded for the film) and original scoring. Both types of music are primarily melodic in their orientation and the cut consistently follows the form of the music, creating a lyrical stretch of film. The differing musics, however, serve conflicting expressive goals, yielding a rather muddled insertion. The filmmakers' desire to both celebrate the helo as a cinematic object, in part by way of new CGI effects, while also setting up the serious and historical battle to come, creates an unresolvable contradiction.

Having arrived in Vietnam—in the narration via an extended musical sequence depicting the officers leaving their families—Moore and his gunnery sergeant share a brief conversation about the coming battle. Both recognize that the odds against them are high: military intelligence has no idea how large the enemy force might be. With quiet resolve despite real apprehension, Moore says, "Let's go do what we came here to do," and the film slips into a musical set piece marked by frequent shifts into diegetic silence. Indeed, on Moore's line a low pedal tone drops into the soundtrack, followed by the entrance of a close-miked male singer with a marked Scottish accent singing the words, "Lay me down in the cold, cold ground / Wherefore many more have gone." The singer is Joe Kilna MacKenzie, member of the self-described tribal band Clann An Drumma based in

Glasgow. Kilna wrote the song, titled "Sgt. MacKenzie," to lament the death of his wife and honor the memory of his great-grandfather, who died in World War I. The song was included on the band's CD *Tried and True*, released in January 2001. Director Randall Wallace heard the recording and brought MacKenzie and the band's bagpipe player into the studio to rerecord the song for *We Were Soldiers*. It's a telling if odd choice. Bagpipe is closely associated with the US military, often heard at military graveside services in reality and on film. Another obvious bagpipe tune that would have been appropriate for *We Were Soldiers* is "Garry Owen," the authentic regimental song for the Seventh Cavalry (Custer's regiment) heard in many Hollywood war films, including *They Died with Their Boots On* (1941), a film about the Battle of the Little Bighorn, Custer's Last Stand; three Westerns directed by John Ford, *Fort Apache* (1948), *She Wore a Yellow Ribbon* (1949), where it's sung by soldiers and used in the score, and *The Searchers* (1956); and in the so-called Vietnam Western *Little Big Man* (1970). "Garry Owen" is a tune with a lineage both military and cinematic. Many critics read *We Were Soldiers* as a return to the war film aesthetics of John Ford. And yet, with an obvious and motivated spot to use it—in a film *about* the Seventh Cavalry in Vietnam—Wallace reached for an adjacent song both less authentic and markedly darker. "Garry Owen," however it's played, is invariably a jaunty marching tune. "Sgt. MacKenzie" is a lament, pure and simple. Here, again, music is a major marker of the difference between war films made before and after Vietnam.

Moore kicks off the mission with the battle cry "Garry Owen," at which point the first group of men board the already-running choppers for insertion. The soundtrack is in a state of near diegetic silence. Muffled effects match each helo as it rises; again the mixer resists completely silencing the image of a helicopter, especially one headed into battle. The music, with the somewhat jarring, unconventional entrance of the vocal, creates the effect of a music video—not for the first or last time in this film. A sequence of thoughtful faces of men headed into battle à la *Hamburger Hill* are intercut with cheering others waiting for subsequent insertion flights.

The music shifts to original scoring by composer Nick Glennie-Smith for the ride in. Voice-over by reporter Joe Galloway, yet to be introduced as a character in the film, locates the moment precisely by date and states that prior to this engagement the soldiers of the United States and the North Vietnamese Army "had never met each other in a major battle." (Indeed, the Vietnam War would not be about "major battles" but rather asymmetrical jungle warfare where the enemy could not be seen and clear battle lines remained undrawn. Most all previous Vietnam PCFs had presented this aspect of the war as it was fought. *We Were Soldiers* centers on just about the only Vietnam battle that can be understood in conventional terms. [*Hamburger Hill* presents both jungle and conventional combat scenes.]) The end of Galloway's short voice-over sees a reassertion of melodic content in the score. A lyrical tune in strings and unseen female voices unspools to the martial accompaniment of a snare-drum cadence beat. The visual highlight of the insertion is a thrilling CGI shot: fourteen helos fly directly toward the camera, passing the viewer's position in a manner no real helos could do safely. Only six real helos are ever shown in a single shot in the film, so the effects shot also augments the number of helos in the film. The next segment again emphasizes the helicopter as a cinematic subject. The lead pilot says—obligingly for the viewer—"drop into nap of the earth." The line of helos proceed to dive and fly close the ground, following the rolling contours of the landscape. Briefly the film indulges the viewer in thrilling movement along the treetops. But then the film's celebration of the helicopter falls away, with a return to more shots of thoughtful men in the choppers preparing for their arrival on the field of battle, no one knowing if the LZ (landing zone) will be "hot" or not.

The thematic confusion of *We Were Soldiers* inheres in sequences such as the above. Wallace wants to have it both ways: to celebrate the sensuous power of the helicopter in the manner of an action film—hence the thrill-ride effect of those choppers flying right at us or dropping to some fancy flying—while also dwelling on the faces of individual men preparing themselves to fight. Eric Lichtenfeld has noted how "the machines [especially helicopters] signify the characters" in action films such as *Rambo III*.[18] Trying to make that equation in a PCF risks losing the subgenre's focus on the fragility of men on the battlefield. In *Hamburger Hill*, and to a limited extent in *Black Hawk Down*, the men remain distinct from the warships they ride into battle. In *We Were Soldiers* the helos as cinematic actors take precedence. But both "Sgt. MacKenzie" and Glennie-Smith's score suggest a fatefulness to the commitment of Moore's men to battle. The helo insertion portends American losses—and there are many—and qualifies any sense of victory the film might offer. In *We Were Soldiers* we do not "win this time." Music, in part, makes this clear before the battle begins.

"Professionalism," the military historian Robert Griffith notes, is "the Army's equivalent of motherhood," to which Beth Bailey adds, "a notion that was virtually impossible to oppose."[19] The SEALs in *Zero Dark Thirty* heading into Pakistan on Stealth Black Hawks to kill Osama bin Laden show no fear or trepidation—it's just another job. Indeed, JSOC commander Admiral William H. McRaven said the Abbottabad mission was "a relatively straightforward raid from JSOC's perspective. We do these ten, twelve, fourteen times a night."[20] The only real complication he noted was doing such a raid in Pakistan. (McRaven speaks here only to the logistics of getting there and back, not the legal issues of a military strike in a sovereign country against which the United States has not declared war.)

Before the raid itself—which is framed by music—the SEALs are given a short stretch of screen time that is itself structured musically. A title card reading "The Canaries" opens the raid section of the film, on which composer Alexandre Desplat activates a rising four-note motif in the horns. The theme plays out in a melodically rounded form over images of the forward operating base (FOB) in Jalalabad, Afghanistan, from whence the raid was launched. In classic PCF form, the noise of helos drowns out the end of the cue. In a rare relaxed moment, the SEALs chat with the CIA agent Maya while playing horseshoes and tossing around a football at their "fire pit"—such spaces show up in all the SEAL films, a kind of elite soldiers' backyard. An indistinct rock track plays diegetically far in the background: "Murder" by the rock group Ours. No viewer is likely to identify the track, although it's clearly not heavy metal or hip-hop. Here Maya gets the call from CIA headquarters that the raid will happen that night. Included in the scene is just about the only joke in the whole film: a SEAL named Justin, played by Chris Pratt, notes he's "cool" with Maya's confidence about bin Laden's location being the only thing between him and "getting ass raped in a Pakistani prison." It's an odd moment and even Maya chuckles, for perhaps the only time in the film. The score restarts as the film turns immediately toward the raid. The command center is shown: this will be a full net sequence, as in *Black Hawk Down*. The SEALs prepare their weapons around a roaring campfire, as if heading out on a hunting trip. The restrained music—Desplat's four-note motif enhanced with rhythmically active violins—is mixed with the clinking sound of weapons being loaded, a typical fetishization of elite soldiers' gear. In a previous scene, Maya spoke dismissively to the men of "your dip and your Velcro and all your gear bullshit." Here, in a brief encounter between the heroine of the film and the warriors who do her bidding, she comes to know and like and even laugh with the combat soldiers who some—perhaps many—in the film's audience have been waiting to take center stage. Images of the men preparing for battle in the firelight are at once mythical and modern: the music gives these moments the aura of a recruiting commercial. This "tool up" scene, like the pre-insertion sequence in *Lone Survivor* to the "Ballad of the Frogman," recalls a cut line from an early draft of *Green Zone*, where "RIPS" of Velcro and clips being slapped into an M-4 are described by one soldier as "the sound of pure sex."[21] The SEALs walk slowly to the helos and climb in: no war cries and running as in *We Were Soldiers*, no defensive crouches against the downwash as in *Hamburger Hill*. As the doors are shut the music cuts off. Before liftoff, the SEALs inside are shown to all be on the net.

The two helos ascend, leaving Maya on the airfield in silence, a familiar trope of the genre. The helos' sound goes to silence almost immediately. The journalist Peter Bergen notes that even stealth Black Hawks are audible "about a minute away from the target," two minutes potentially if the wind affects conditions.[22] The sound of the helos in *Zero Dark Thirty*—featured in a shift to an effects-only mix—fairly evaporates into the dark night. Maya looks after them; in the distance all she sees is an American flag waving over the base. The moment chimes, in a cinematically telescoped time frame, with General McCrystal's memory, quoted in the epigraph to this chapter, of the chorus of helos dissipating "until silence returned to the darkened base."[23]

Maya does not go on the raid—although she monitors the comms from the command center. Most viewers likely know that President Barack Obama, members of the cabinet, and Pentagon brass watched the actual raid from an improvised setup at the White House. Then Secretary of State Hillary Clinton commented, "This was like any episode of *24* or any movie you could ever imagine."[24] The audience for *Zero Dark Thirty*, of course, gets to ride along. The raid itself was discussed in the previous chapter. Here, I turn toward the musical aesthetics of the helo insertion and exfil. Bigelow and Desplat combine to remake this PCF trope for the JSOC military of the GWOT.

The ABA—music, joke, music—sequence around the fire pit does important work in adjusting the viewer's expectations. Here, *Zero Dark Thirty* moves into combat mode proper, with a brief period of rest before the action begins. The tense quality of the musical frame sets off the relaxed atmosphere among the men. The SEALs are not afraid. They are a bunch of guys at a barbeque waiting for the call to go to work. When preparing their weapons and boarding the choppers, they show no emotion. Indeed, they are presented in silhouette or near total darkness. If they had emotions on their faces, we would be unable to read them. Nor does the music seem to feed on their preparatory energy: the score is too dark and active, and they are all cool and economical in every way. (In a detail matching Jonathan Pieslak's research into how soldiers use music, SEAL Matt Bissonnette, a participant in the raid who published an account under the pseudonym Mark Owen, recalled "the thundering beat of a metal band blaring out of some speakers" at the "fire pit" just before the actual raid.[25] In line with the GWOT cycle's approach to popular culture and elite soldiers, *Zero Dark Thirty* leaves out the metal.) These moments before the ride into Pakistan set up a distinction between the SEALs in the film and the American viewing audience. *Zero Dark Thirty* is about to step off into its moment of truth. The audience should be excited; the SEALs should not.

Zero Dark Thirty cuts from Maya at the silent FOB to the two helos in flight. A grand and dramatic musical cue begins, placed first on the film's score album, the track called "Flight to Compound." A slow, low beat sets the expansive scale of the music, with nervous, pulsing figures above in brass and vibraphone. Designating specific instrumentation is difficult, as the mix of the music suggests heavy processing and an electronica vibe. This is not the sound of a conventional concert orchestra—even though the London Symphony Orchestra is credited in the film. The music fills the time it takes to fly from Jalalabad to Abbottabad, using a musical interlude to suggest distance traveled. The SEALs wait patiently. Justin provides another light moment. When asked what he's listening to on his earbuds, he says "Tony Robbins," the motivational speaker and author. Justin tells his buddies he wants to share what he's learning and says, "I got plans for after this." A few SEALs chuckle. But what is the audience supposed to do? The lightness of the chatter—so different from all previous helicopter insertions in PCFs—speaks to the routine nature of JSOC's war. Bissonnette writes of the pressure he felt internally to make the raid just another day's work: "What we were about to do was significant. We fought hard to keep history out of our minds."[26] *Zero Dark Thirty* will have none of this. The SEALs are all confidence. The only other spoken exchange comes when a SEAL asks the men for a show of hands as to who has been in a helo crash before: all respond affirmatively. This exchange is, again, for the viewer—preparing us for the first event at the compound, when a helo does, indeed, crash. For the SEALs, the ride to the compound is akin to a commute. For the viewer, it is much more—and so Desplat's score represents their feelings of excitement. After all, there is no surprise to *Zero Dark Thirty*'s denouement. All that matters is the how of what happened. As noted, the actual mission is presented in a mix of realistic and limited perspective, all on the side of the SEALs. The lack of music during the raid works to keep emotion at a distance. Desplat's "Flight to Compound" takes a different tack, investing the insertion with a restrained grandeur but keeping the SEALs themselves at a remove. This music is for us—not them. The viewer is not asked by the music to identify with the SEALs. Instead we are given time to enjoy the dark grandeur of two Black Hawks in flight through the dark mountains—American military muscle and tax dollars at last exacting payback for 9/11.

The only real event during the insertion comes when the helos cross into Pakistani airspace, announced to all on the net, heard by the viewer as voice-over. At that moment, Desplat taps old orientalist tropes: a slightly exotic descending tune comes into the score in a quavery mix of woodwind and strings—again the precise instrumentation is hard to parse. Here—and only here—do the SEALs seem to get serious faces. None, however, show the trepidation and fear common to insertion scenes in earlier films. Rather, it's a moment of putting on the game face, not a screaming war face but a serious, get-the-job-done look. At the comms' voice-over "three minutes" they begin final preparations, donning their night-vision goggles—the most expensive kind, with four lenses. (Bigelow establishes the green vision perspective used in the raid here as unattached to any single SEAL.) The music tails out as they approach the target and is gone for the duration of the raid.

Music returns only for the very end of the mission: as the downed helo is destroyed and the remaining helo departs. It's the same music as "Flight to Compound." Much time is spent on the demolition of the downed helo, which provides a movie-worthy explosion to cap off the raid. An administration official watching in Washington noted of the explosion as seen on the drone video feed, it was "like a Jerry Bruckheimer movie."[27] Bigelow uses this generically apt actual event strategically—combat sequences have almost always ended with a big explosion, a punctuation displaying American firepower and articulating a structural point in the narration as well.[28] The explosion is shown four times from four perspectives: it's a constructed action-movie moment that clears the soundtrack.[29] What's left is the four-note motive, in a new extended version, first in all strings, then a lone trumpet with an erhu in canon, playing over a greatly tightened return journey to Jalalabad, where Maya is waiting. The viewer is given only a short amount of time to decompress: Desplat's score is, yet again, for them, not the SEALs. Indeed, back at Jalalabad the only SEAL to show any emotion is, once more, the voluble Justin: bin Laden's body deposited at the FOB, he removes his helmet and shouts a single "whooo." No other SEAL joins him. All in a day's work.

MUSIC AS ATTACK "PATH"

The helicopter attack in *Apocalypse Now* by the First of the Ninth Air Cavalry to the strains of Richard Wagner's "Ride of the Valkyries"—led by Lieutenant Colonel Kilgore, played by Robert Duvall—is one of the most famous action sequences in movie history. John Milius's original 1969 script includes the attack, with the Wagner, and describes the helos in motion as "almost a dance of dragonflies."[30] Like the helicopter insertions described above, music structures the representation of helicopters on-screen—only here the music also accompanies actual combat. Using upward of ten helicopters provided by the Philippine air force, director Francis Ford Coppola simulated an airborne attack on a Vietnamese seaside village with astonishing verisimilitude. Movie audiences had seen restaged air combat in the 1969 film *Battle of Britain*, which used orchestral music to accompany British and German fighter planes dueling high in the sky. The difference in *Apocalypse Now* is the helicopter: a vehicle that moves close enough to the ground to register relative speed and to allow the battle to include humans, both attacked by and combating the "dragonflies" of war. The combat climax of the Kilgore sequence comes when a helicopter pursues and men on board shoot from the air a Vietnamese woman running away from an exploding medevac helo that she destroyed with a grenade—the earliest Hollywood spectacle of a US helicopter going up in flames. The asymmetrical nature of the Vietnam War and the ability and will of the Vietnamese to fight the seemingly overwhelming technological force of the US military are captured in a single, tightly told, explosively visual vignette, ending a long stretch of almost unbearably exciting filmmaking.

The lingering power of the Kilgore sequence is strong. Jonathan Pieslak interviewed a soldier during the Iraq invasion who related how his unit used the music alone to refer to the film: "During a 'thunder run' into Baghdad, [he] said they blasted Wagner's 'Ride' on the outside of their truck as they attacked, as in the movie." The soldier articulated two reasons for doing this: first, "the music was motivational for the American soldiers who knew the scene from the film"; second, they assumed that Saddam Hussein, a movie buff, knew the scene as well and so his troops would be fearful and surrender. Here, American soldiers reenact *Apocalypse Now* by playing the soundtrack only—a borrowed piece of classical music that has no connotations for them outside Coppola's film—as a means "to hasten the Iraqi forces' surrender through psychological intimidation" and to make their own war making into a kind of homage to a film they all knew and assumed the enemy knew, too.[31] As noted earlier, *Jarhead* (2005) the film includes Marines watching *Apocalypse Now* together before deploying for Operation Desert Storm in a large auditorium, where its effect would be enhanced. *Jarhead* and *Apocalypse Now* were both edited by Walter Murch. In both films, Murch uses the Wagner to make an enduringly disturbing point about combat: it is sensually and sexually exciting.

Murch makes this point by rendering combat as also musically exciting—so exciting, it induces the Marines in *Jarhead* to sing along. In the most telling encounter between the two films, the Marines watching and singing continue Wagner's melodic line past the point when it falls silent in *Apocalypse Now* on a cut to the village before the attack. The moment brings Wagner's music to the fore as the organizing force for the entire scene. This helicopter attack is indeed a dance, or perhaps, more specifically, an opera. In the film's press pack, Coppola said *Apocalypse Now* was "intended as a film opera." This aesthetic goal reaches fruition in Kilgore's attack, which musicalizes the movement of helicopters through space and locks the machines into the soldiers' identities with gleeful, sexual directness built directly on Wagner's musical language.

Murch has described how the sequence was shot and edited:

> The helicopter attack . . . was staged as an actual event and consequently filmed as a documentary rather than a series of specially composed shots. . . . Once Francis said, "Action," the filming resembled actual combat: Eight cameras turning simultaneously (some on the ground and some in helicopters) each loaded with a thousand-foot (eleven-minute) roll of film. At the end of one of these shots, unless there had been an obvious problem, the camera positions were changed and the whole thing was repeated. Then repeated again, and then again . . . each take generating something like 8,000 feet (an hour and a half).

Murch characterized the very slow process of turning this mountain of footage into a unified sequence as "not so much a *putting together* as [the] *discovery of a path*."[32] Much of that path was provided by Wagner's "Ride of the Valkyries" as recorded by conductor Georg Solti leading the Vienna Philharmonic Orchestra in 1966 for Decca Records. Coppola watched rushes to the sound of this recording in Manila. Murch assembled the sequence in California using Solti's recording as well. With much work already done using the Solti, it proved difficult to get permission to use the recording in the film. Decca initially refused. Murch tried multiple other recorded versions but none satisfied him at the level of tempo fluctuations—an obvious and partly practical issue for a sequence cut so closely to preexisting music—and the quality of orchestral sound. As Murch noted, "One particular shot that remains vivid in my mind—looking down from one of the helicopters, past one of the bombs, you can see the ocean. In Solti's recording he had chosen to emphasize the brass at that precise moment and there was something about the acidity of that blue, combined with the metallic nature of the sound, that was synergistic—the blue of the ocean and the brass that had been emphasized, fed upon each other."[33] Trying again to obtain the clearance, Decca allowed *Apocalypse Now* coproducer Tom Sternberg to ask Solti personally. The conductor granted permission.

The form and content of Wagner's music as interpreted by Solti shapes the editing of the attack, laying out the timing of the sequence and the course of events and, most importantly, glossing the attack and the men who do it with specific musical meaning. Coppola and Murch's edit to the music links the often-fierce pleasures of Wagner's musical arrivals to the act of firing fearsome weapons at the enemy. The viewer's doubled pleasure in the music and the attack is further enhanced by the sonically incredible conceit that Kilgore's men are themselves hearing the music, that they are taking pleasure in firing on the beat. The quasi-sexual release of combat—a common trope in many combat memoirs—here achieves cinematic realization in a deeply disturbing sequence that weds the helicopter as an instrument of war to accompaniment drawn from the canon of Western art music. Milius must have heard an aesthetic resonance between the Wagner and the hard-rock tracks originally slated to serve as Kurtz's diegetic war music.

The Kilgore sequence begins with Wagner's music being sung into the film by an unseen soldier who overhears Kilgore assuring Willard that he can set the patrol boat down anywhere they need it. Kilgore says, "This is the first of the ninth, air cav, son, airmobile," to which a soldier responds by heartily singing a snatch of Wagner's "Ride." The close link between going on the attack and musical performance is established here. On the remixed *Apocalypse Now* double album, the singer, joined by others, makes it to the end of the first phrase of the Wagner, reinforcing the point in a way only hinted at by the film's mix.

As Kilgore gives orders for the coming battle, the whine of helo rotors starts up on the soundtrack—a vaguely synthesized sound, rising in pitch and intensity to back up Kilgore's shouted punch line, "Charlie don't surf." On a direct cut to the next morning, the chop of many rotors bursts abruptly forth in a roar of noise. Preparing to take off, Kilgore engages in sonically unrealistic dialogue near his helo, casually chatting with his door gunner, utterly oblivious to the roar surrounding them. Imperviousness to the dangers of modern combat—sonic and physical—proves a keynote to Kilgore's character. It is first established here. In an even more unbelievable exchange of words, Kilgore tells a bugler, standing at the ready near the chopper, "let 'er rip." The bugler proceeds to blow the charge, his sound rising clear in the morning light as the many helos depart—an effect only possible in the synthetic sonic realm of a film soundtrack, where individual sounds can be modulated up or down in a way they cannot in any real acoustic context, an act Murch describes as "playing the mixing board the way you would play a piano."[34] This musical send-off to the helos—added to the script by Coppola—announces that what follows will be sonically unbelievable.[35] As is so often the case in *Apocalypse Now*, the mix, carefully balancing effects and music, offers an interpretation of the images as a kind of madness rather than an authentic documentation of the war, even though this moment and the attack that follows offer an immersive experience of filmed simulated actual combat, made possible by the unique resources of the film.

The conversation in Kilgore's lead helo sets up the attack to Wagner. Coppola asks the viewer to believe, quite literally, that helicopters and music go together. The Wagner, we learn from Kilgore, serves as a form of "psy-war ops," played loud both to scare the "slopes" and rile up his men. In an article for the website psywarrior.com, (retired) Sergeant Major Herbert A. Friedman documents the use of loudspeakers mounted in helicopters in Vietnam and quotes pilot First Lieutenant William Tyner recalling psy-war missions in 1968 using music to accompany the dropping of leaflets:

> When we had a dual role mission the pilot had to operate the cassette while maintaining an orbit somewhere near the target area. As the leaflets were drifting in, we would drop down as low as 1500 ft (over safe targets) and play the propaganda tape. "Rock and Roll" recorded off of AFVN was my favorite and seemed to attract the most ground fire. "Fire" by Robert Brown was the clear winner. Mostly though, the tapes were professionally made discordant funeral music, wailing women and rude comments meant to ruin morale, comments like: "You didn't see that B-52 coming did you?"[36]

In practice, psy-war missions were separate from attacks and involved single helos only. The music played from helos targeted the Vietnamese only, often in culturally specific ways to which the Americans hearing this music would not have responded. Had Milius and Coppola represented actual psy-war practices using music and loudspeakers, the audience for *Apocalypse Now* would not have responded, either.

Kilgore initiates the attack with the words "Romeo Foxtrot, shall we dance," and presses play on a reel-to-reel tape machine. A sequence of rhyming shots follow, each zooming in on a large speaker mounted in the door of a different chopper. The zooms are timed precisely to the series of trills that open the Wagner, establishing what will be an almost point-for-point scaffolding of film form on musical form. The speakers are pointed toward the ground, yet the men *in* the helos behave as if they can hear the music, despite the deafening rotors roaring just above their heads. The final script describes the moment the tape starts: "Suddenly we HEAR the

strange, AMPLIFIED SOUNDS of the WAGNER. All the men in the copter's faces light up."[37] The viewer capable of thinking realistically at this
point—a resistant, overliteral moviegoer, assuredly—knows that this is
sonically impossible. The technical issue of coordinating the inputs into all
those speakers before wireless technology alone defies belief. Indeed, the
1975 script nods toward technical issues: "INT. SPEAKER-COPTERS.
Blasting out the Wagner. One is having difficulty making the tape work.
The MUSIC STOPS and STARTS as the recording is wound by finger."[38]
What really defies belief is the conceit that the music coming out of all
those speakers, however loud, could bounce back toward the helos from a
great height off the surface of the sea and compete with the roaring rotors
and play in meaningful synchronization. And yet the soundscape, all out of
scale with our experience of the actual world, is irresistible. This is insist-
ently diegetic music—our visual attention is drawn to the sound sources
repeatedly—and it's crucial to the scene as edited that the men in the choppers also hear the music. As discussed earlier, Milius's script uses the verb
blare in close association with rock music Kurtz plays at his compound. The
capacity of the Wagner to convincingly "BLARE" is important. Still, the
conceit of the music as diegetic and heard by all is frankly ridiculous.

Wagner accompanies the flotilla's flight for several phrases, while the
image track cuts between grand vistas of the helos and shots of the men,
some of whom are preparing their weapons. The 1975 script describes these
moments: "The men—nervous, excited—very few of them really scared—
they fondle their rifles, grenade launchers, claymore mines, anti-personnel
grenades, plastic explosive cord, flame throwers, M-60 machine guns,
expendable rocket launchers, mortars and bayonets."[39] One soldier ritualistically taps a magazine of cartridges on his helmet before attaching it to his
rifle—an act re-performed with a box of candy by a Marine watching
Apocalypse Now in *Jarhead*. The action is given substance by a prominent
tapping sound effect that, again, defies the sonic environment. Several cuts
during this stretch of music celebrating the men and their machines fall on
points of melodic arrival in the music. The enjoyment of the music as a
force coordinating the choppers and the men and their combined imminent
attack forms part of the pleasures of this pre-battle sequence. This section
is perhaps the one that offended John Irvin, who accompanied similar shots
with "We Gotta Get Outta This Place" in *Hamburger Hill*.

But just before a prominent musical climax, the film cuts abruptly away
to the village about to be attacked, where all is quiet. (In *Jarhead*, the
Marines watching and singing along keep singing at this point, finishing
the phrase that Coppola and Murch deny the quiet viewer.) Briefly, before

the sounds of choppers and Wagner signal the coming Americans, the schoolchildren are heard singing. Wagner disrupts the songs of village life: the battle is, in part, a musical one. A lone Vietnamese boy in the village looks toward the audibly approaching squadron instead of running to safety. A cut to the helicopters advancing on the village, presumably what the boy sees in the distance, matches the first entrance of a singer in the Wagner. Enhanced prominence of the music in the mix lends the voice a predatory edge. The choppers and a screeching soprano are both headed the boy's way. A direct cut to a pilot's-eye view, also on a phrase arrival and featuring another small bump in the music's volume, answers our desire to see this scene from the American side—the "side" of the music—and disturbingly trades our previous identification with the boy for a view to a kill.

Cuts between village and helos continue, at least one shot—a sweeping helo moving below the camera—matching exactly an upward-sweeping figure in the music. The musical tension builds toward resolution as the helos close on the village. The imminence of musical arrival communicates the final seconds before the battle begins. Milius's script includes command language starting the battle: a pilot dispassionately measures the approach with the words "700 yards—600—500—commence firing."[40] The music—measuring on a more subliminal and pleasurably anticipatory level the ground being covered between the choppers and the village—makes such dialogue superfluous. The start of the battle is fundamentally musical. The first rocket is fired on the close of an elided phrase, an arrival that lets both the gunships and the low brass rip just as the soprano sings a long, ecstatic high B. For the viewer attending to the music's unfolding, the act of firing that rocket can be anticipated well before it happens (particularly on repeat viewing): it's a controlled release that hits its mark exactly in the music. But any viewer familiar with Western music—anyone conditioned by Hollywood movie music—will feel the move toward firing the first rocket. We hear the moment coming and can enjoy the orgasmic pleasure of anticipated release yielding explosive results.

A full attack on the village ensues, playing though Wagner's complete chorus-like statement of his theme. But Coppola and Murch required more music than Wagner provided, for after the initial musical climax of the "Ride," marked by three female voices singing a sustained, triumphant chord, the cue cuts seamlessly back to earlier in the recording, extending the duration of the music to fit the desired length of the battle. In other words, Coppola and Murch recut preexisting music to match the needs of the film: it's a song scene in which the recording organizing the action is remade. The edit of the images and combat sound effects helps mask this abrupt rewind to an earlier point in the music: a thrillingly fast POV shot from a helo sweeps over the village square on the vocal climax, visually matching the music, then close shots of door gunners firing on automatic fill the soundtrack with the noise of weaponry, momentarily drowning out the music and allowing for a return to earlier in the track (around where the first singer enters). After this, the buildup in the music, already used once to structure the firing of the first rockets, accompanies and structures a narrative vignette turning on the destruction of an antiaircraft gun on the beach. The music proceeds toward the vocal climax—already heard once—but this time deployed to bring the Wagner to a close. The final vocal trio chord is matched to a flying shot moving through a black cloud of smoke. Having passed through this barrier, a sequence with strictly diegetic sound follows with tactical attention to the details of the battle communicated by various pilots and Kilgore on the net.

A final musical sequence remains: the Wagner returns—with no apparent diegetic source—for helos landing on the beach delivering men to the battlefield. This music cuts out abruptly, and the Wagner exits for good, when an American soldier is shot in the village square, the first casualty pictured. The destruction of the medevac and shooting of the Vietnamese woman who does it follows, as does the downing of a second helo. These events—neither given musical shape—bring the action sequence to a close.

Setting the helo to music in *Apocalypse Now* and in several other sequences described above expresses a kind of wishful thinking that the helicopter is something it is not: suppressing the sonic roar of the rotors, literally musicalizing the machine in active combat, changing the helo's essential nature. And while Coppola and Milius's sonic conceit may be absurd, it is not abused. Kilgore does not win the day without sustaining losses. And his losses are not musicalized but instead accompanied by the noises of battle and disturbing chatter on the comms—as in the improvised line, "Kill that dink bitch." In that respect, the Wagner heard by all in the "first of the ninth" represents the imagined efficacy of the helicopter in the minds of these men—and perhaps, momentarily, those watching them. The cinema, of course, enables such fantasies. But even *Apocalypse Now* ultimately calls them into question.

PART IV

Music

8. Unmetered

Before Vietnam, Hollywood war movies marched.

Movies about the Marines almost invariably used "The Marines' Hymn," a smart, once universally known march, as primary musical material. Such movie score musical branding lent authentic tuneful luster to what Lawrence Suid has called "the one branch that over the years has best publicized its role in the nation's martial history."[1] The "Hymn" dominates Marine films made during and just after World War II—for instance in *Wake Island* (1942), *Guadalcanal Diary* (1943), and *Sands of Iwo Jima* (1949)—and resounds into the 1950s. Max Steiner's score for *Battle Cry* (1955) uses select phrases at key moments, such as when the men press on after the death of a beloved officer, and David Buttolph's less subtle score for *The D.I.* (1957) uses it constantly. In this historical context, the almost complete absence of the "Marine's Hymn" in Marine-centered PCFs is remarkable. In these films, it's heard once (often ironically) or not at all. *The Boys in Company C* (1978) begins with the "Hymn" heard against the "boys'" less-than-heroic farewells to their families. Near the end of *Full Metal Jacket's* (1987) first half, the "Hymn" plays for the boot camp graduation parade, with Hartman's final harangue spoken in voice-over: "Marines die. That's what we're here for." (At film's end, Stanley Kubrick's Marines go marching on to their own singing of the "Mickey Mouse Club March.") The Marines in *Windtalkers* (2002) sing it at a bar—a less than convincing moment. A more typical, sardonic, and very brief use of a singing soldier has Captain America, a ridiculous and incompetent officer in *Generation Kill* (2008), briefly humming the "Hymn" in an early episode, cementing thereby his distance from the Recon Marines at the heart of the story. The "Hymn" never once sounds in *Born on the Fourth of July* (1989), which favors "Grand Old Flag," nor in *Flags of Our Fathers* (2006), the latter a rather astonishing omission, as the film re-creates one of the Corps' most celebrated battles as well as the dedication of the Marine Corps War Memorial in Arlington National Cemetery. Finally, *The Pacific* (2010), an epic, ten-hour story of three Marines, never once sounds the "Hymn," not even at John Basilone's Medal of Honor ceremony. The musical link between the Marines and "The Marines' Hymn" has been broken in popular culture. (My students typically cannot identify the tune.)

When not handed a real military march as their main theme, composers for combat films made before and during Vietnam almost to a man wrote jaunty original marches. *Battleground* (1949) offers an immediate postwar example. Many entries in the post-1960 cycle of epic battle re-creation films, such as *The Longest Day* (1962) and *Midway* (1976, score by John Williams), audibly evoke the cocky use of the "Colonel Bogey March" in *The Bridge on the River Kwai* (1957), complete with whistling. The "Colonel Bogey March" even became a pop hit. Soldiers singing marches in rousing unison turn up even in more reflective films such as *A Walk in the Sun* (1945) and *Men in War* (1957)—the latter has a quiet, chamber music–style score for most of its duration. And singing men make a late appearance in the march-style theme for John Wayne's *The Green Berets* (1968), a film whose score, as Wesley O'Brien notes, would not have been out of place a quarter century earlier.[2] "The Ballad of the Green Berets" was also a hit record, reportedly blasted from college dorm rooms by unsympathetic students to sonically disrupt antiwar demonstrations.[3] A late-1960s cycle of "dirty" war films, so described by Jeanine Basinger, often combine the impudent march with rock beats and pop timbres (as in *The Dirty Dozen* [1967] and *Kelly's Heroes* [1970]), lending a hip edge to frequently less-than-heroic accounts of battlefield exploits. On television, the assertive, aggressive march theme for the long-running series *Combat!* (1962–67) included sound effects of explosions detonating on the beat. All these marches impart a jaunty confidence and masculine bravado to the American fighting man.

In *The Warrior Image* (2008), Andrew J. Huebner argues for a slow and steady transformation of popular perceptions of the American soldier from 1941 to 1978. "The terms of what made a soldier honorable—indeed, the terms of what made him a *man*—were widening and changing" across these years, Huebner writes, but "the agony of combat for the American foot soldiers, increasingly apparent throughout the 1950s and 1960s, burst into full view when large numbers of people began questioning the value of [one] *particular* war"—Vietnam. Looking at various media and PCFs from the late 1970s only, Huebner concludes, "Imaginative representations of the Vietnam War also *reinforced* and *amplified* many ideas about war ascendant since the late 1940s, suggesting that perhaps the difference between Vietnam-era imagery and what came before was a matter of degree more than essence."[4] Huebner does at times comment on music in media representations of the American soldier. He notes, for example, how scores for military-produced Korean War news films remained "jaunty," "upbeat," and "chipper" despite depictions of soldiers as exhausted or in deep sorrow.[5] But music in the PCF, from the start of the subgenre, marks an abrupt departure from earlier "jaunty" practices. While other aspects of the representation of soldiers might have changed by "degrees," in the expressive realm of film scoring, the advent of the PCF initiated a change in "essence."

The remainder of this book turns toward the musical element of the soundtrack. Chapters 8, 9, and 10 define new musical tropes of the PCF. Chapter 8 considers veil musics, often used to characterize the exotic other encountered by soldiers fighting in foreign lands. The word *veil* refers to a common approach to texture: a light, transparent, sustained background against which isolated sounds in irregular rhythmic patterns are heard. Chapter 9 looks at alternatives to the momentum-giving march, including waltz-time (or triple-meter) themes for films set during World War II and driving electronica beats for films set in the Middle East. The elegiac register, a subgenre-defining musical innovation that runs across the full history of the PCF, is described in chapter 10. The final chapter considers music serving specific formal functions: at the ends of PCF narratives and during the final credits.

VEIL MUSIC DEFINED

Composer Dick Halligan's score for *Go Tell the Spartans* (1978) captures in innovative musical terms the complicated situation of US troops in Vietnam. The first sound in the film is a lonely, rather weakly played snare drum offering a spare quasi-military cadence, too slow and too rhythmically ambiguous to coordinate marching men. This American military sound—a typically powerful instrument played in equivocal fashion—is immediately answered by metallic, vaguely "Asian" sounds, "exotic" percussion in an even more obtuse, implied triplet rhythm. Further layers are added to each of the initial percussion matrices, each distinct and suspended in a texture at once thick with sound and transparent to the ear. On the "Asian" side, high strings play sliding figures, evocative of the erhu, and an "ethnic" wood flute moves about with its own musical ideas. On the American side, subdued horns (or low-range trumpets) play fragmented pieces of a melody that never add up to a whole. The snare contributes a roll every once in a while, but no cadence ever solidifies. Sometimes the horn melody and the strings slide to a shared point of arrival but there is no way to predict where this music will go. Indeed, it seems to be more a stacking of sounds in space than a movement forward along melodic or even basic rhythmic lines. This mix of contradictory timbres and rhythmic patterns suggests stalemate from the start.

The combination of contrasting musical elements introduced during *Go Tell the Spartans*'s main titles continues across the film's narrative of an American outpost surrounded and overrun by the Viet Cong. The only musical elements not introduced at the outset are an echoing fanfare—perhaps a nod toward Jerry Goldsmith's score for *Patton* (1970)—and an "ethnic" flute melody that plays when a young soldier encounters an old Vietnamese man (perhaps a ghost) struggling to defend his homeland. Most of Halligan's score feels, in context, as if without meter, as if any musical event might happen at any moment, any sonic element enter at any time. The score's only metered cue puts a military drum cadence to ironic use, mocking a newly promoted, pathetically unqualified American lieutenant addressing his Vietnamese "troops"—in fact, poorly armed farmers who laugh in incomprehension at his speech. The organizing principle of the march has been lost and cannot be regained in this tale of American defeat. Over the final credits, Halligan tosses out lonely fragments of the familiar tune "When Johnny Comes Marching Home." This ghostly evocation of a marching song reads as mournful and forlorn rather than ironic or sarcastic, of a piece with the earnest tone of *Go Tell the Spartans* and most PCFs. The score, a matter of fragments, enacts in musical terms the loss of national direction implicit in the loss of march time and, by extension, the US defeat in Vietnam.

Halligan's score for *Go Tell the Spartans* was the first to define a musical register heard across the PCF, a equivocal kind of combat movie music grouped here under the term *veil music*. The major difference between Halligan's score and subsequent veil music is the former's inclusion of traditional American military sounds: snare drum and brass. From *Platoon* (1986) onward, veil music has served mostly as a means to characterize the foreign "other" encountered by American soldiers on overseas battlegrounds. As such, veil music can be heard to express the experience of asymmetrical warfare from the US side—the enemy other as a mystery (*Platoon, Black Hawk Down* [2001]). That said, veil music has proven to be flexible in its meanings: it can be turned toward a sympathetic view of the foreign other (*Casualties of War* [1989], *Three Kings* [1999]) or, with the removal of exotically marked elements, toward representation of more general states of psychological distress or combat tension (*Full Metal Jacket*, *The Hurt Locker* [2008]). It is also appealingly ambiguous in its meanings (again *Platoon* and *Black Hawk Down*). Before considering how veil cues function within specific PCFs, the musical qualities of this innovative register warrant general description.

Veil cues are characterized by a generally soft to moderate dynamic level. This is a quiet sort of movie music, not appropriate for noisy battle scenes but well suited to the tense moments prior to a firefight. Low volume makes veil cues flexible in how they enter and exit the mix, sneaking in or tailing out with ease. But veil music is of sufficient heft to make a statement if it falls suddenly silent. While the veil metaphor suggests sonic transparency, the register is thick with timbral variety, with a generally light texture made up of many contrasting, often quite resonant layers of sounds. A visual analogy clarifies this defining textural aspect. Veil cues are marked by a sustained sonic background: a ground against which other sounds, or figures, are heard. This ground can be composed of high strings, synthesized sounds, or "nonmusical" sounds bordering on noise or static. Tone clusters—dissonant bunches of notes rather than traditional dissonant chords—are common. Melodramatic seventh chords, which might suggest good guys in distress or the fundamentally evil nature of bad guys, are absent. Blurring the line between acoustic instruments and synthesized sounds also plays a role in making veil cues ambiguous. Music can sound unmoored from conventional meanings when its point of origin is not clear to the ear. *The Hurt Locker* pushes this effect to an end point of sorts. Such sonic grounds do not lead the ear in any predictable direction, nor do they shape the durational experience of the music. Rather they set an all-around tone, functioning like atmospheric sound effects suggesting wind, insects, or waves. However, while such sound effects might serve to establish location—for instance the jungle—veil music indicates a kind of affective temperature. An array of shorter, more pronounced sounds act as figures set against—or perhaps behind or within—the transparent "veil" of the sustained background. Specific instruments and vocal styles mark different veil musics as representative of different wars. Sounds otherwise classifiable as effects can also play the role of figures. Veil music figures occur in irregular rhythmic patterns. This allows veil cues to last as long as needed and to efficiently and subtly catch points of action, underlining the formal nature of the register as a matter of texture and timbre rather than meter or melody. Most veil cues do not slip into any kind of beat, and extended or closed melodic shapes are rare. Veil cues do not impose their shape or duration on the image track as, for example, strongly melodic themes do. When a veil cue does acquire a pronounced beat or recognizable melody, the register is being turned toward a specific purpose (as described below in the case of *Casualties of War*). It's a slippery, unpredictable sort of movie music, open to expressing suspense, fear, dread, mystery, or just a sense of general otherness (relative to the default Americanness of the PCF).

VEILS IN VIETNAM

Veil cues are heard in most all Vietnam PCFs of the 1980s. All but one of these scores (*Full Metal Jacket*) rely on traditional orchestral instruments. The backgrounds to veil cues in *Platoon, Casualties of War*, and *Born on the Fourth of July* employ techniques and textures drawn from modernist classical music of the Western tradition. Sources include the so-called night music of Béla Bartók, evoked by Ennio Morricone in *Casualties of War*, and the modernist techniques of György Ligeti and Krzysztof Penderecki, tapped by John Williams in *Born on the Fourth of July*. Georges Delerue's score for *Platoon* draws on both sources. Set against these backgrounds are audibly "Asian" figures: clanging cymbals or gongs ring out, overblown wood flutes burst forth, bright wood sticks click in their own overgenerous acoustic, nasal string instruments slide between pitches in mini glissandos.

Delerue's most original contribution to war film scores comes with a series of cues written for moments of extreme uncertainty in the jungles of *Platoon*. Delerue creates a music of maximum menace out of quiet but resonant materials. These register-defining cues omit any musical reference to US forces: there are no snare drums or brass as in Halligan's music for *Go Tell the Spartans*. The first veil cue in *Platoon* begins when Chris first sees the enemy—in silhouette, in the dark of night—while keeping watch. A high string pedal comes in on an effects mix noisy with jungle insects, demonstrating the ease with which an obviously musical sound—strings—can enter on an active mix of diegetic effects. The jungle sounds continue even as the string pedal splits into a dissonant cluster—still high strings only—which changes unpredictably and without harmonic direction. Added to this mix of music and diegetic effects is the decidedly subjective sound of a heartbeat, presumably Chris's own. At a moment of great tension, Stone has the soundtrack lay out three distinct kinds of sound: veil cues, designed like background effects, are well suited to the creation of complex sonic structures that combine music and effects. This transparent but thick texture persists for many seconds—drawing out the moment and building tremendous tension. The score and heartbeat—the latter clearly not a musical sound—grow slightly in volume, moving toward the moment when the tension is released by an offscreen explosion, the start of the battle, which restores the soundtrack to a completely diegetic mix.

Delerue's use of sustained clusters is echoed in Williams's cues for both of Kovic's battlefield mistakes in *Born on the Fourth of July*: he and his unit fire on a hut full of civilians, and he shoots and kills an American soldier in the heat of battle. In both cases, Williams, like Delerue in the above moment from *Platoon*, draws on modernist techniques found in the 1960 and 1970s compositions of Ligeti and Penderecki: clusters shifting in an unmetered context, and harsh or buzzy timbres and textures. Modernist techniques such as these were being heard more and more in Hollywood scores: Stanley Kubrick interpolated pieces by Ligeti and Penderecki in both *2001: A Space Odyssey* (1968) and *The Shining* (1980), and Williams used these techniques for the alien abduction of a small boy in *Close Encounters of the Third Kind* (1977) and in some of his score for *E.T. the Extra-Terrestrial* (1982). Their application to the combat genre was new.

Later veil cues in *Platoon* add exotic elements that suggest the otherness of the Vietnamese enemy: specifically a low-range flute playing a fragmented melodic gesture (which sounds fragmented because it does not evoke Western notions of musical closure); a churning, wrenching, sliding figure deep in the low register; and extremely resonant wood sticks clicking at unpredictable intervals, which seem to resound in a different acoustic from the rest of the score. The studio-made mix of this music is crucial. Veil cues, even when using sounds drawn from acoustic instruments that the viewer can identify, process those sounds in a confected, unnatural acoustic generated in the abstract sonic space of the recording studio. The space-making quality of film scoring—inserting a musical zone or filter between the viewer and the diegetic world of the film—allows for such synthetic acoustics to intrude upon without destabilizing the realm of the film, provided diegetic sound remains in the mix. (Usually tied to potential danger, veil cues seldom operate in diegetic silence.) When applied to the very "real" space of a combat zone, the effect can be uncanny. The longest stretch of this sort of veil music in *Platoon* begins when Sal picks up a booby-trapped box. Sal's arms are blown off: a slightly slow-motion image accompanied by soft but menacing veil music. The music continues as the platoon struggles to respond to invisible enemies who are, apparently, all around them. It plays through the discovery of the grotesquely garroted Manny, and Barnes's snarling delivery of the line "the motherfuckers," only ending when Barnes shoots a Vietnamese villager running in the distance. Barnes's shot initiates the village atrocity scene, the troubling center point of the narrative. Here, a complex, shifting, entirely unmetered veil cue gives musical substance to a general feeling of fear and vengeance, marking the stunning success of the enemy at invisibly ambushing the US soldiers and cementing the resolve of the Americans to respond. In frustration, they wreak havoc on innocents—albeit in a village where they discover stores of rice and weapons.

Similar veil music returns when Barnes shoots Elias and when Chris shoots Barnes. The flexibility of the register is on display here—as in the scene described above. Veil cues in *Platoon* are morally neutral. They are not solely identified with the Vietnamese other, and their inherent unpredictability and menace can just as easily underline an American killing another American. As action-movie music that allows for a range of ambiguous meanings, Delerue's veil music signals a heightened state of possibility. Anything can happen when this music is playing. Indeed, the most intense moment of moral decision in the film is scored with veil music. After the final battle, Chris finds the gravely wounded Barnes crawling on the ground. A veil cue—complete with low flute and resonant bright sticks—fades in just before Barnes is revealed. Chris is already stalking him. Barnes says, "Get me a medic," and Chris raises his gun—an AK-47, the enemy's weapon. Barnes taunts him, saying, "Do it." Chris shoots: the shots, again, cutting off the music. Delerue's original cue for this moment reacts strongly to Chris's shots with high, dissonant, and loud string clusters, which dissipate and fall, eventually yielding to a mournful statement of Delerue's elegiac main theme (modeled closely on Samuel Barber's *Adagio for Strings*, last discussed in chapter 2). This cue—rerecorded in 1995 on a CD containing the rejected score for *Platoon*—lasts almost a minute longer than the cue in the finished film.[6] Its content, shifting from the menacing veil music to the mournful elegy theme, suggests a missing segment of the film during which Chris showed immediate remorse for or just reflected on the act of killing Barnes. At some point in the process of editing *Platoon*, Stone gave the viewer time to reflect on Chris's act, and Delerue scored it with elegiac music. This time to reflect was later removed. In the finished film, a modest musical tag, still in the veil register but much less active, plays while Chris walks away from Barnes's body. Within seconds, an approaching vehicle's motor clears the soundtrack to diegetic sound only.[7]

Veil music is flexible and unpredictable, but the slightest suggestion of either melody or meter can tip it toward specific thematic ends. In *Casualties of War* Morricone uses the elements of veil music found in *Platoon*, with the exclusion of the resonant bright sticks. But the veil music, like the film itself, is on the side of the other: Oanh, the victim of US soldiers who either actively harm her or fail to save her. The majority of Morricone's score uses a veil-like texture activated melodically and rhythmically by a flute melody that cycles again and again, tracing out Oanh's travail and effectively locking in the viewer on her side. This music is first heard during one of Brian De Palma's signature POV Steadicam shots. The POV is Meserve and Clark stealing into a village to kidnap a girl. With a tremolo or synth strings background, two melodic ideas are introduced: a plucked instrument (likely a guitar but without Western cultural connotations) plays what will be Oanh's theme, while a pan flute bursts forth unpredictably with a second figure, also to be associated with Oanh. At the moment of choosing her—Clark calls her "the pretty one," and Meserve responds, "Take the pretty one"—the music panics along with Oanh, growing instantly more urgent. In retrospect, the music represents Oanh's fear, while the image gives us the gaze of her tormentors. These two melodies—both modal, both prone to circular repetition—follow Oanh throughout her travail up until the moment of her death.

As always, *Full Metal Jacket* offers a unique angle on a PCF trope. Composer Vivian Kubrick (billed as Abigail Mead) uses a version of veil music at key moments of violence, inflecting the register with a menace that draws on synthesized sounds and a physically palpable, "breathing" meter. Kubrick composed and performed the score on a Fairlight music computer, in an effort, she said, "to avoid any prior musical associations that conventional orchestra instruments might bring."[8] The goal of diminishing conventional associations is squarely in line with veil music across the PCF, an important part of its investment in otherness and unpredictability. The wood sticks of *Platoon* are especially prevalent here, made even brighter in the entirely synthetic mix. The creepily resonant sticks unfold along their own pattern-less way, bearing no apparent relation to the other elements. Kubrick's veil cue has an unmistakable in and out rhythm: a meter akin to the slow, heavy breathing of a monstrous metallic beast. Christine Lee Gengaro hears "the sound of a respirator in a hospital," another quasi-mechanical metaphor.[9] A slow-moving, irregular, cycling melodic figure comes and goes but reaches no point of pathos (as Oanh's passionate and deeply human flute melody does). This music is heard on three occasions in the film: when the men of the company beat Leonard with bars of soap wrapped in towels (the so-called towel party); when Leonard shoots Hartman and kills himself; and when Joker kills the injured female Vietnamese sniper. All of these occasions show characters deciding to use violent, even deadly, force. The calculated choice to harm is made in a context of heightened sonic stakes created by a menacing veil music that signals the possibility that anything might happen.

VEILS IN THE MIDDLE EAST

The score for *The Hurt Locker* can be fruitfully understood as mostly veil music. As noted in chapter 2, the film's score—with sufficient recognizably musical content to warrant an Oscar nomination—is a collaborative creation between the music and the sound departments. Director Kathryn Bigelow described this tension: "When I say sound, that's all of it, that's all the effects, all the detail work, and of course the score, because all of it needs to feel like it's one piece. It has to be able to be integrated and yet set itself apart at the same time so it's a complicated proposition."[10] Bigelow's desire for the score to be both integrated and set apart pinpoints the flexibility of veil music—especially when both ground and figures draw on electronic sources rather than more specifically "othered" musical timbres (such as the gongs and wood blocks of Vietnam veils). Across the length of *The Hurt Locker*, sonic textures that signal something "score-like" sneak into an otherwise very active effects mix. These veil cues are disarmingly subtle—you have to listen for them—and, as with earlier veil musics, they serve several functions. One exceptional veil cue acts quite traditionally, resolving, as it were, on a point of action-derived discovery, a rare moment of Mickey Mousing in this otherwise-diffuse score. On his first mission in the film, Sergeant James uncovers an IED (improvised explosive device) that turns out to be a group of shells daisy chained together in a wagon-wheel arrangement. Just after James declares "we're done," the score indicates that danger remains with a low pedal that enters over cuts between James and an Iraqi man watching from a balcony above. As James follows a wire, the sonic texture—still all synth strings—thickens and adds dissonance. The cluster that accumulates grows in volume and narrows to a single pitch as James discovers the extent of the bomb yet to be defused, exactly matching his action of pulling the center of the daisy chain. Co-composer Buck Sanders recalled Bigelow requesting that the score "sting the moment" with a "spine tingling feeling." Co-composer Marco Beltrami noted they took a "Penderecki type of approach."[11] It's applied here with maximum efficiency to musically gloss a visually arresting moment that ended up on the film's US poster. The cue continues as James defuses the bombs. His snipping of the many wires and the echoing steps of the Iraqi man, fleeing the scene down a stairwell, work as effects-derived figures. This sequence proves an exception in *The Hurt Locker*. Most of the score works more subtly, blending into and becoming part of the effects soundscape. Beltrami noted one concern of the creative team, that "the danger of using music might take the viewer out of the picture if it was scored in a traditional sense."[12] The utility of veil music as atmospheric sound, closely wedded to sound effects, that does not automatically activate the viewer's musical sensibilities is crucial here. Veil music offers scoring that works toward audience immersion. The rhythmic flexibility of veil music contributes as well: as co-composer Sanders notes, "You never know what's around the next corner, so musically it shouldn't be predictive." He continues: "Since it's not thematic music, but more of a sonic based score with textures and ambiences, we reuse those styles a lot because it's like you're repeating certain themes for a film that has a variety of themes in the score."[13]

Among these "textures and ambiences" are the varied contributions of Yorgos Adamis, who Beltrami called a "Greek guy" who played an array of "different various ethnic flutes" and offered "vocalizations [with] a Middle Eastern sound to them."[14] Adamis's contributions, prominent in the first minute of *The Hurt Locker*, are best heard as Middle Eastern–ish—a decidedly vague musical "other" that proves typical of the register as applied to PCFs set in the Muslim world. Veil musics heard in PCFs set in the Middle East regularly include othered voices singing in a foreign language, their words never translated, remaining opaque to the American audience. Only rarely is this voice legible as the Muslim call to prayer. Instead a more generalized, often quite abstractly othered, kind of veil music predominates, generally avoiding characterization of the Middle Eastern other as inherently evil (at least in the score), opening these films to many interpretations and also serving as decorative elements suggestive of an "oriental" brand of exoticism. Here, as in the Vietnam cycle, veil music expresses something of the unknowable other—specifically a language-based mystery. As Dexter Filkins writes of the Iraq War, "The most basic barrier was language itself. Very few of the Americans in Iraq, whether soldiers or diplomats or newspaper reporters, could speak more than a few words of Arabic. A remarkable number of them didn't even have translators. That meant that for many Iraqis, the typical nineteen-year-old army corporal from South Dakota was not a youthful innocent carrying America's goodwill: he was a terrifying combination of firepower and ignorance."[15] Ignorance proves an aesthetically useful aspect of the untranslated voices heard in many PCFs set in the Greater Middle East.

Three Kings and *Black Hawk Down* demonstrate how veil cues can flexibly if imprecisely fix the identity of the foreign other while also serving to point the viewer toward the American soldiers at the center of the story.

Composer Carter Burwell's score for *Three Kings* abounds in active, often chaotic percussion grooves: out-of-meter drum solos, vaguely suggesting jazz, usually underscoring the rogue US soldiers' improvised attempt to steal Saddam Hussein's gold. But at several key junctures, veil cues come into play. Three are worth analyzing for their contribution to heightening combat moments and their construction of both the Iraqi other—in this case, a sympathetic other—and the American soldiers. During the village shoot-out, a synth-strings background sneaks in just after Troy, played by Mark Wahlberg, is shot. The isolated instances of firing provide a percussive element. The music responds as a kind of expanding halo, (literally) adding resonance to the soundtrack with each shot and investing each subsequent bullet fired with additional importance, as if the acts of violence—each graphically portrayed in slow motion—are building to an aggregate importance that demands sonic acknowledgement. This veil cue unobtrusively contributes a weight to the air, scoring without excessive melodrama Troy slowly opening the Velcro closure on his Kevlar vest and pulling off a flattened slug, an action marked by a vibrating bell-like sound in the score. Figural elements in veil cues can easily act as subtle stingers. Soon thereafter, the American "kings" and the Iraqi civilians flee together in a jeep and a commandeered truck full of the gold. The Iraqi army, back at the village, launches a gas attack by way of mortars that finds the fleeing party, their vehicles disabled, in the middle of a minefield. The explosion of the first gas-filled shell kicks off a sustained passage of dense veil music. Here, the mix of figural elements is active and urgent. Exploding shells, heard but not seen, add an effects-derived element of chaos to the music. A more emotionally weighted veil cue comes at the close of the film, when the three Americans reveal to their superiors that they have the gold in exchange for ensuring their new Iraqi friends get safely across the border into Iran. The primary figural element is an edgy, nasal-voiced, "exotic" solo string, which plays a tunelike melody lacking clear meter. Helo rotors far back in the mix keep the diegetic world of the film engaged. A randomly pinging, softly metallic sound plays the part wood sticks might take in Vietnam. The cue exemplifies the dramatic possibilities of veil music as unmetered, sonically sustained background that lends an emotional temperature to a scene and can follow events unobtrusively. Once the plan for the refugees is set, a hoarse, youthful, Arab male voice enters singing a wordless wail filled—in context—with extreme emotion. In this case, the voice of the "other" works to express the inner emotions of the Americans, who have (apparently) put aside their greed to help the Iraqi refugees (who, the plot suggests, have been otherwise abandoned by the US military and government). The voice sings for the American trio, accompanying their reaction shots as they watch the Iraqis escape. But who's to say what the voice means on its own terms? The provenance of this othered voice is completely unknown. To the Hollywood movie ear it registers as both deeply expressive and definitely other, as undeniably human yet not American. The ease with which this expressive voice blends with conventional Hollywood scoring is demonstrated as the refugees wave goodbye to their American saviors. Studio orchestra strings come in with a schmaltzy musical theme—unlike any heard so far in this self-consciously hip film—that, like the dramatic situation itself, would not be out of place in a feel-good, made-for-television movie. The Arab voice sings on with the strings, suggesting the union of the Iraqis and the American soldiers who, in contrast to official US policy, embody an honorable response to the Iraqi people post–Gulf War. Veil music, which thrives on the simultaneous playing of disparate sounds, opens an easy path to musical expression of this moment.

The opening of *Black Hawk Down* lays out several musical figures set within and against a continuous synth strings and wind effects background. These individual elements—none set to a clear meter—come forward singly or in combination, only to pile up at the close of the opening when the film is released into its beat-driven narrative score. Here the relative "otherness" of individual figural elements is, again, more evoked than nailed down, reliant on an implied American moviegoer's sense for what is "ours" and what is not. The first section represents the suffering of the Somali people. A low pedal tone, hissing and humming, establishes the cue's ground. The first figural element is another buzzy, non-Western stringed instrument, establishing the foreign locale of the film as somewhere in the Middle East. This element continues behind the next figure: an untranslated male voice, singing in a non-Western language enters with dynamic power as the image track opens on the sight of a terribly thin body being wrapped in grave clothes by a man with his back to the camera. The singer is Baaba Maal, a Senegalese vocalist singing in the Pulaar language in a style native to his West African home. Senegal, a sub-Saharan nation, lies as far to the west of Somalia as it is possible to get before hitting the Atlantic Ocean. While Islam is the majority religion in both Senegal and Somalia, the ethnic makeup of these two nations is utterly different. For the American viewer, Maal's voice registers as generically African "world music": audibly *not* Western and, crucially for this moment, not connoting Islam. Maal's solo continues over images and texts describing Somalia's civil war and famine, offering a highly rhetorical wail linked visually to the sorrow and hunger of Somalia but completely inscrutable at the level of actual content. Few American viewers will know Maal's music or the tradition it comes from, hence there is no way for the film's target audience to assess how accurately (or not) this music matches the images and text it accompanies. Instead, Maal's vocal can only be understood by Hollywood listeners as within a context of otherness and, given the perceived power and conviction of his performance, as an embodiment of evident strength and passion. To repeat, there is no way to know what this music means within its own tradition: it could be a love song, it could be a comic song, it could be anything. As such, it cannot be heard as counterscoring (anempathically)—hearing a contrast between image and music requires that the viewer know what a given music means conventionally within its own system of meaning—nor can it supply added specific meaning to the image. In *Black Hawk Down*, Maal's untranslated vocal, in combination with the images and titles, comes off as a song of sorrow. Hans Zimmer's score hijacks the raw material of Senegalese music and assigns it as it pleases, in this case as a generically African—or at least non-Western, nonwhite— sound, giving voice to stylized, stereotypical images of African famine.

The figural elements in Zimmer's veil cue shift on the title card "the world responds." With the background synthy pedal tones and wind effects continuing, a slowly oscillating melodic figure in high synth strings harmony repeats several times, never in a strict meter—like a great heaving sigh (not entirely unlike the breathing meter of Kubrick's veil music in *Full Metal Jacket*, only here without menace). After the passion of Maal's vocal, this expressively restrained, higher, recognizably Western sound comes as a relief, matching the cup of water offered by a white aid worker to a thirsty African. The buzzy string instrument keeps coming and going as well: this music, like all veil music, is changeable and many layered. A final figural element—a sonic representation of a helicopter rotor, not quite a sound effect but audibly a helo, in Michel Chion's term a "rendering" of helo sound by or as a musical sound—enters with a title card noting the arrival of "America's elite soldiers" in Mogadishu. The helo rises in prominence and begins to set a beat. The buzzy string solo finds a descending contour and—at long last—the music finds a point of clear arrival. The beat drops, catching an action-film-style cut to a title card reading *Black Hawk Down*. Don't worry, Zimmer's score assures, this is an action film. And indeed, the veil music used to lay out the backstory for the film is rare in the score, which is almost entirely beat driven (as discussed in the next chapter).

What does veil music do at the start of *Black Hawk Down*? First, it establishes a context of complete otherness. Non-English vocals are disturbing to American audiences: the American popular music marketplace is famously resistant to untranslated vocals, and foreign singers who want to enter the US market sing in English. Opening the film with the extreme otherness of Baaba Maal immediately locates *Black Hawk Down* at a far remove from home. Second, veil music economically, if from a slanted angle, presents the power equation in Somalia without granting the Somali warlords any musical presence: only the suffering Somalis, helpful Western aid workers, and elite US soldiers with powerful flying machines are heard from. The combat scenario of the plot will prove quite different: crazed crowds of Somalis bringing down Black Hawks to the sound of powerful and complex beats. Third, this veil music crucially omits any element that might connote Muslim identity. Somalia is not presented as a Muslim country, an omission that again marks *Black Hawk Down* as a pre-GWOT representation of the region.

Zimmer's opening-titles music recalls Halligan's mix of Asian and American sounds at the start of *Go Tell the Spartans*. In both cases, the coming into contact of two cultures is represented in characteristic sounds. The difference is Zimmer's use of suggestive helo noise rather than snare and brass to represent the United States and the way the *Black Hawk Down* titles close with a launching of the score's beat. *Go Tell the Spartans* never finds the beat. Both films tell tales of a US defeat, but *Black Hawk Down's* score strives to secure a sense of focused American action that *Go Tell the Spartans* simply does not pretend is possible.

The last extended veil cue in *Black Hawk Down* accompanies the so-called Mogadishu Mile. While evacuating to the safety of a soccer stadium, a group of US soldiers are left in the dust by the convoy of Humvees behind which they were running. Exposed to the last embers of the long firefight, they engage in a literal running battle. Slowly the mix begins to strip out diegetic sound as low humming pedal tones, an Arab melody, tremolos, an intermittent heartbeat figure, and Maal's voice combine in a veil texture of almost dreamlike stillness. The mix of elements creates a state of extreme exhaustion, but it's unclear what the connotations of any given sound might be. Unlike the opening music—where Maal's voice was tied to the suffering Somalis and helo sounds indicated the US soldiers—the sensuous veil cue for the Mogadishu Mile is, perhaps, simply decorative. It gives the surviving heroes of the film a last chance to show their mettle: they run in slow motion, heightening the suggestion of extreme exhaustion. Some stumble and fall, others fall while trying to help. They enter a bank of fog or smoke and diegetic sounds sneak in. Soon a young Somali boy is running with them, and crowds of smiling and waving Somalis are lined along the street. They reach the finish line—entering the soccer stadium—in a kind of triumph. (The sequence is a fantasy. In the actual battle, no such reception by the Somalis occurred. *Black Hawk Down* casts Somali extras as stand-ins for its American audience, cheering on the heroes in the home stretch.) Arriving in the stadium, the men are greeted by Pakistani waiters with glasses of water and confronted by the bodies of their dead comrades being zipped into body bags. The veil texture shifts to high strings only as all the foreign figural elements fall away. This extended section—American soldiers looking on each other and their own dead—forms a lyrical coda to the story of the film, a needed decompression, time to take stock by looking only, not talking or analyzing. And not analyzing is key: if the PCF is to center on the experience of soldiers, then after-battle analysis is beside the point. Who lived and who died is all that matters. Sweet and sad, "pure" synth strings veil music, activated easily from the previous more timbrally diverse and exotic veil texture, aptly accompanies the long, reflective stretch of time we share with the soldiers who survive the Mogadishu Mile.

Early in *Black Hawk Down* there is, however, a nod toward Islam. In a beautiful flying shot marking the start of the day of the raid, the camera circles around a huge minaret. Two voices—again singing in a foreign tongue and style—sing against a background pedal tone and the sound of ocean waves. Added to this mix is the call of the muezzin, seen on the minaret's balcony. Here a veil-like texture is created by a combination of scoring, sound effects, and diegetic sung speech. The singers in the score are likely Lisa Gerrard and Denez Prigent. Later in *Black Hawk Down* the pair sing a duet titled "Gortoz a Ran" in Breton, another language American audiences are unlikely to identify, carrying with it an kind of exotic frisson but, in fact, stunningly distant from Somalia in origin. Prigent is a Breton folk singer and songwriter. Gerrard, a frequent Zimmer collaborator, is an experimental world music vocalist who often sings in made-up or private languages. Mixing the voices of Prigent and Gerrard with the muezzin aids in creating a kind of sensual mix in the soundtrack. This score can seemingly absorb anything and, indeed, veil cues welcome a tremendous variety and novelty of sounds. The moment is curious in ideological terms as well: the shot ends on a man who bows in prayer, then picks up an AK-47 with his prayer mat and walks off. This man, perhaps marked by his weapon as an enemy other, will shortly be revealed as the Somali informant who tells the US command which house to raid in search of the warlords they hope to capture. The capacity of veil cues to provide simplistic sonic exoticism should not be discounted. Moments such as this, however they play out plot-wise, help establish a context of pervasive otherness within which the Americans—soldiers and viewers together—must find their way.

9. Metered

Meter serves an organizing function in music and culture. William McNeill's classic 1995 study *Keeping Together in Time: Dance and Drill in Human History* shows how central the military march was to the creation of standing armies and the nation-states raised by and made with them.[1] War films lacking marches tacitly admit a failure of the nation-state to (literally) keep together. Confronted with US defeat in Vietnam and a country marked by military failure, PCF makers have drawn on various other strongly metered musics: waltzes, electronic dance beats, stately ostinatos. This chapter considers two general approaches to meter, each characteristic of a particular PCF cycle: first, the use of triple meter in the World War II cycle, and second, beat-driven scores in twenty-first-century PCFs. While similar musical approaches to scoring are highlighted in both these cycles, the expressive power of composers' choices always inheres in the specifics of a given film. As with veil cues, similarly metered cues prove flexible in their meaning. The chapter concludes with a close reading of the central battle episode in Terrence Malick's *The Thin Red Line* (1998), the "Attack on the Bivouac," set to an unchanging, strongly metrical cue. This thoroughly musicalized combat sequence, much of which unfolds in near or complete diegetic silence, offers an extreme counterexample to the effects-driven battles described in chapter 6. In Malick's battle, music by Hans Zimmer drowns out most of the blasts and screams of a thoroughly ambiguous stretch of combat action, challenging the viewer to find a stable position on the spectrum between immersion and reflection.

The larger issues connecting all the films discussed in this chapter are as much structural as expressive: How do original scores displaying a high degree of metrical uniformity support and shape larger historical themes, cycle-specific combat actions, and viewer engagement? The answers to such questions in films that do deploy marches are rather straightforward and limited. Traditional military music is not subtle. As used in war films, it has but two registers: march on proudly to victory, or march on with a jaunty air (perhaps suggesting some impudent resistance to the strictures of military life) that in the end also arrives at victory. Reacting to the loss of Vietnam, PCF scores use different strongly metered musics to tell markedly different stories.

WORLD WAR II IN TRIPLE METER

The long-lived World War II PCF cycle sets aside both marches and swing. What's left? In three prominent cases, the answer is the waltz. *The Thin Red Line, Band of Brothers* (2001), and *Flags of Our Fathers* (2006) alike employ triple-meter themes in a moderate waltz-like tempo. The latter two share a sentimental transgenerational agenda linking fathers and sons, and the waltz—a dance music historically gendered female—works in both as unlikely musical support to these efforts. *The Thin Red Line* explores the existential state of soldiers somewhat suspended in time and history and employs triple meter to suggest a kind of floating equilibrium, where thoughts of life and death swim about in soldiers' minds. Below, I first compare the triple-meter themes in these three PCFs, then chart how those themes are employed structurally to narrative and thematic ends. In all three cases, triple-meter melodies emerge as musical distillations of these films' larger ideological projects: either framing the old for the understanding of the young, or presenting the dilemmas of soldiers as tied up in philosophical questions cutting to the heart of the human experience.

Seven writers, seven directors, and two cinematographers divided the duties of writing, directing, and shooting *Band of Brothers*. Only composer Michael Kamen's creative voice runs across the entire series, unifying the whole by means of a medium—music—that specializes in expressing what cannot be put into words. *Band of Brothers*'s main theme suggests how precisely Kamen controls the emotional work done by the score. The theme is entirely diatonic: every pitch in the tune is in the F major scale, a compositional choice that yields a deceptively simple, perhaps even naive, tune. Dissonance—nudging the theme toward elegiac pathos—is created by means of pedal tones in the bass and repeated stepwise descending sigh figures in the melody that give this generally harmonious music an underlying hint of longing. The moderately paced theme played by an orchestra of mostly strings has a ceremonial, public quality, indicating the collective address of the series to the nation as a representation of the nation's fighting men. But Kamen's melody is audibly waltz-like—it's not too slow to dance to—and retains a certain tenderness, however grand the arrangement. Dance music, after all, is at once public and private: the waltz is a partner dance, its rather intimate embraces sanctioned by the social space of the ballroom. Like many waltzes, Kamen's theme can be brought into the home. It works nicely as a piano solo and has been marketed as such; more than a few amateur pianists offer their versions on YouTube. Subtly mixed, wordless choral voices doubling the strings on the tune—perhaps an echo of Williams's similarly scored "Hymn to the Fallen"—encourage the listener to understand the theme as song. It's easy to fall into the habit of singing or humming along with the theme on its return at the start and finish of each episode, especially when binge watching *Band of Brothers* with friends or family. This aspect of Kamen's theme has the potential to bond viewers to each other in the act of impromptu musical performance done *with* the show while also allowing each individual to supply their own meaning to a tune that comes with no lyrics attached.[2]

Clint Eastwood composed a monothematic score for *Flags of Our Fathers*. The film's only melody is a diatonic tune, usually heard in triple meter, not particularly vocal given its many leaps. The tune is spare in its design, allowing for melodic embellishments, some filling in the long notes with graceful figures. The tune also includes a final melodic coda, similar to the short phrase tacked onto the end of Kamen's theme for *Band of Brothers*. Both waltzes have closing phrases that signal the end while also extending the musical form. In Eastwood's case, the coda is, at times, repeated and drawn out. His theme can be arranged to suggest musical closure as imminent but, for now, put off—until it does come to final resolution. In this respect Eastwood's waltz is not dramatic but meditative and, importantly, finished—if at times reluctantly. It cannot be quoted in part but must be played in full, a formal requirement with implications for its use as narrative scoring.

The Thin Red Line overflows with music in triple meter for strings. Zimmer's equivocal, at times tentative score offers a cluster of related triple-meter melodies heard variously on ten occasions, some for extended stretches, including at the film's close. The group of themes are similar enough to warrant discussion as a larger aesthetic: all but one move at about sixty beats per minute, all offer rounded but not particularly tuneful or memorable melodies—one reason they blend into each other. In addition, an oscillating vamp figure, also in triple and sometimes incorporating harp, occasionally precedes or follows these melodies as introduction or coda, further blurring the distinctiveness of the individual tunes. There is, in short, a slow, triple-meter register to Zimmer's score, one element of the entire film's moderate, restrained musical content. (The score album track "Light" combines many of these melodies into a kind of triple-meter medley.)

And so, while Kamen and Eastwood's themes come to clear, emphatic cadences, Zimmer's cluster of themes does not. The former pair offer consistent and satisfying closure the latter generally avoids. Kamen and Eastwood's triple-meter themes are audibly dancelike, activating reminiscence of the waltz more directly than Zimmer's meandering melodies. Since around the turn of the twentieth century, the waltz has served as a reservoir of nostalgia: old-fashioned, sentimental, feminine, domestic. It carried these qualities even in the 1940s, when most American soldiers were fans of big-band swing, a driving four-beat style of popular music. Kamen's decision to make the score for *Band of Brothers* waltz instead of march or swing—a choice no doubt made in consultation with co–executive producers Steven Spielberg and Tom Hanks—adds a precisely calibrated sentimental element to the series' telling of the combat journey of Easy Company. Eastwood seems to have arrived at triple meter in his exploratory path toward a theme for the film. *Band of Brothers* and *Flags of Our Fathers* use their waltz themes to similar formal and expressive ends, if in ways that reflect their different media contexts.

For all its cinematic production values, in formal terms *Band of Brothers* remains a television show, and so the musical theme that sticks most persistently in the ear is that heard over the opening title sequence—a formal element of many television series that proves important in *Band of Brothers*. Kamen's waltz-time main theme is heard at the start of each episode and during each episode's lengthy closing credits, which scroll in the manner of a feature film. In both sequences, music alone fills the soundtrack, in spite of several explosive, by implication noisy, images seen in the opening credits. This music—which critics heard as "lyrical" and an example of musical "ennoblement"—provides an overarching emotional continuity to *Band of Brothers*, literally framing each episode in waltz time.[3] (As I discuss at length elsewhere, the main theme is also used for dramatic scoring across the series.)[4]

Visually, *Band of Brothers*'s opening titles function as a stylized scrapbook of images culled from the series. The viewer catches glimpses of battle but spends most of the titles peering into the faces of young men. Crucially, the meaning of the images in the opening titles changes as the series progresses. What begins as an assemblage of faces reacting to sights unseen gradually morphs into a memory book of important moments from the series captured in evocative stills. As the episodes go by, more and more of the credit sequence images offer reminders of scenes already seen. The credits also keep alive the faces of characters who die or depart from the front line before the close of the series. Kamen's hummable theme accompanies this entire process, always present to usher the committed viewer back into the community of Easy Company. Preserving signal scenes in a sort of musical amber—reflected visually in the distressed or antiqued images, many presented in a bluish wash—the main theme, heard musically whole, orders the characters and events of *Band of Brothers* under the sign of triple-meter motion that commits to no specific historical time period—this is not music of the 1940s—and that avoids excessive or melodramatic sentimentality by means of harmonic and melodic restraint.

Every episode except the last begins with documentary-style interviews with Easy Company veterans. The old men's words are never given musical accompaniment at the top of the show, increasing the contrast with the musically rich main titles sequence. Only at the close of episode 10 are the veterans named—allowing the viewer to match each old man with his younger counterpart in the narrative—and only here does music play behind their words. The score withholds music from the old faces until the very end when, at last, they too are bathed in the waltz that has repeatedly supported the actors playing their parts. Not until our final encounter with the veterans do we—the viewers—have the feeling of coming fully to know them, matching their stories as embodied by young actors in the series to their actual old faces and voices, all to the sound of a melody ingrained in our ears over many hours spent watching *Band of Brothers*. This moment of sentimental recognition—finally knowing who's who—is crucially supported by Kamen's music. The film historian Thomas Schatz has described the inclusion of the veterans as "at once personalizing the narrative and injecting a sense of documentary realism," adding, "as the series wears on, both the aged veterans and their dramatic 'characters' become increasingly familiar, and in an odd sense the older and younger versions of the Easy Company warriors gradually fuse."[5] Sustained visual attention given to old and young faces at the start of each episode, and Kamen's sentimental waltz—nostalgic, by generic association, on first hearing, and specifically nostalgic to *Band of Brothers* after repeat hearings across the series—work a kind of alchemy in this "fusing" process.

In the opening moments of *Flags of Our Fathers*, just after Eastwood is heard singing a phrase of "I'll Walk Alone" unaccompanied in the blackness, the score opens with the main theme played by solo piano against a high string pedal. In voice-over an older, male voice—unidentified at this point and never clearly locked into a speaker—warns the viewer not to trust the easy black and white of war stories: "Heroes and villains. Most of the time they are not who we think they are." The resolutely modest scale of the "heroes" of the film is established here in words and music. (The voice-over works as an entrée to Eastwood's two films taken as a pair: *Letters from Iwo Jima* [2006] offers a "not who we think they are" lesson about the Pacific War's "villains.")

The triple-meter tune on piano, with a simple accompanying line for the left hand, returns at the film's halfway point. After the Iwo Jima landing, the young corpsman Doc Bradley walks a wounded soldier to an aid station near the tide. The theme plays in a rounded form over images of the monumental task of those working just behind the front line, including soldiers doing graves registration. Doc's perspective on these sights is emphasized for about half the sequence but several shots lose him on his way to the aid station, offering an unidentified POV on the scene. The viewer, in effect, takes a walk with and not with Doc, as he and the unmoored camera take time to reflect on the scene. Bradley's walk ends with him standing before a row of dead bodies, which stretch to the horizon, the tune serving as a lullaby for sleepers who will never awaken. When the melody ends, Eastwood cuts to a Japanese artillery piece high on Mount Suribachi. It fires, and the film drops right into a running battle. The use of an explosive combat event to kick off a shift in the mix from music to effects, and in the narrative from reflection to action, could hardly be more bluntly done.

In the film's final half hour, Eastwood's main theme—heard only three times in the first hour and a half—begins to dominate the soundtrack, tying together a series of scenes that bring the narrative to closure. The main theme plays during twelve of the final twenty-three minutes (not including the lengthy final credits), accompanying dialogue, voice-over, and action. While the music for this long, slow, final movement is not continuous, repeated returns of the main theme—it makes eight prominent entrances—unify the closing sequences and, in particular, the story of Corporal Ira Hayes, a Native American Marine, one among the men who raised the flag. The theme plays when Hayes confesses his fear that Sergeant Mike Strank, who dies on Iwo Jima, would not be proud of him, and again when Hayes leaves the war bond tour after showing up drunk too many times. The theme's coda, reminiscent of the end of a slow movement of a Mozart piano concerto with its slow harmonic rhythm, sustained strings, and limpid right-hand melody, fits well with the image of Hayes's train pulling out of the station. A guitar version, with a bit of twang, accompanies Hayes's walk from Arizona to Texas to tell the family of Corporal Harlon Block that their son was—as Mrs. Block believed—among the men in the photo, and also plays over Hayes's final days as a field laborer and his death from alcoholism. When the old John Bradley dies in a hospital bed— his son who researched the story of the flag raisers at his side—Eastwood's theme is heard for the only time in a strings-only, contrapuntal texture that evokes the elegiac register derived from Samuel Barber's *Adagio for Strings* as used in *Platoon* (1986). With the final voice-over—an epigraph spoken by Bradley *fils*—the piano returns, playing the main theme as at the film's opening. Bradley counsels the viewer in voice-over, "If we wish to truly honor these men, we should remember them the way they really were, the way my dad remembered them." The elder Bradley dies just after recalling to his son how, after planting the flag on Mount Suribachi, the soldiers went swimming. The narrative of *Flags of Our Fathers* ends on the image of young soldiers on the beach at Iwo Jima. Scoring this scene, and so much of *Flags of Our Fathers*, with a delicate, lightly orchestrated waltz keeps the film firmly focused on the personal. The cliché of the final voice-over— "They may have fought for their country, but they died for their friends"— comes off as devoid of masculine posturing in this context where a persistent and modest, generically feminine, musical theme has reduced the legacy of the war to the personal—in the end, finding its deepest meaning in the relationship between a father and son.

Like almost all the many themes in the score for *The Thin Red Line*, Zimmer's triple-meter theme group is applied freely and never linked to any single character. At various points it is tied to Bell and his wife—including playing behind remembered physical encounters between the pair and during her "Dear John" letter heard as voice-over. A long cue in triple, the only one paced above 75 bpm (beats per minute), plays when Captain Staros departs the film, telling his men in voice-over that they are his sons. On several occasions ruminative triple meter is attached to the graceful, gentle Witt, particularly in his role as a foil for the cynical, unbelieving Welsh. Triple meter first enters just after Welsh and Witt's first of three talks. In the brig, reflecting on their conversation, Witt strikes a match in the darkness and utters the line, "I love Charlie Company." Witt will, at film's end, lay down his life for Charlie Company: Welsh will look upon Witt's grave to the sound of a triple-meter tune and ask, "Where's your spark now?" The notion of a spark within men was the subject of Witt and Welsh's final conversation. Welsh asks Witt how he is able to believe, saying, "You're a magician to me." Witt replies, just as the by now very familiar triple-meter vamp begins, "I still see a spark in you." The film follows this statement, made to rather equivocal music, with a complex visual-sonic conceit, all to the sound of triple-meter melodies. Welsh is shown walking through the grassy slopes among his men. The unassigned Southern voice-over (described in chapter 5) begins to speak about death, noting, "Death's got the final word" but that while one man only sees the pain, another man "feels the glory." On the word *glory* Welsh looks down at Witt sleeping peacefully in the grass. The film's final voice-over, spoken by the never-identified Southern soldier, is also accompanied by a triple-meter tune. Triple-meter music supports several of Witt's ruminative voice-overs, in particular just after the first battle when he attends to wounded men by a river and wonders, "Maybe all men got one big soul." The triple themes also play while Witt comforts the dying Keck ("You didn't let your buddies down," he tells the man who fell on his own grenade) and when he decides to sacrifice himself for Charlie Company by staying behind to distract a group of advancing Japanese (the Christlike Witt tells Fife, "You go on ahead," reminiscent of Elias's line "I'll haul it for you" said to Chris early in *Platoon*). Malick's construction of a kind of sensitive soldier's oversoul—the combination of the unassigned Southern voice asking existential questions and Witt's words and actions—is crucially supported by Zimmer's triple-meter themes, which cycle, recycle, and vamp across the length of the film, setting a moderate pulse that doesn't dance so much as flow. Zimmer's cluster of similar melodies combine in a murmuring, slightly sad but sometimes hopeful, ever unrolling melodic stream—an analogue for the ruminations, spoken and thought, of the men who move through Malick's Guadalcanal. The lack of clear closure and specific tuneful identity to Zimmer's triple-meter melodies opens a musical space within which the collective stream of consciousness of the film's tangle of voice-overs finds a place to dwell. In this formal respect, the expressive conceits of *The Thin Red Line* rely on the suspended stasis of music in waltz time.

BEAT DROPS IN TWENTY-FIRST-CENTURY PCFS

> The radio a soundtrack that adrenaline has pushed into silence,
> replacing it with a heartbeat.
> —BRIAN TURNER, "2000 lbs." (2005)

GWOT PCFs often foreground the heartbeats of twenty-first-century warriors in both the narrative and the mix. Beneath their gear, the "cool professionalism" of elite contemporary warriors resides in a vital and warm, sounding human body.[6] *Act of Valor* (2012) and *Lone Survivor* (2013) alike foreground a SEAL heartbeat to clock a soldier's survival: marking the limits of underlying human frailty, charting a "never give up" spirit. In *Act of Valor* the hearts of the LT and the Chief beat as one for a brief moment. The

Chief is gravely wounded at the end of the long final battle. In a context of subjective sound, we hear his heart beating. The film transitions to the mortally wounded body of the LT, who had thrown himself on a grenade. The heartbeat transfers to him and, in a long, slow zoom in on his unblinking eye, his heartbeat slows, then stops. The LT's death—like that of the cherry Gardner in *Platoon*—is a sonic event. In *Lone Survivor* Marcus Luttrell's heart rate is monitored by the medics reviving him after his rescue in a scene that opens and closes the narrative. At the close, when it seems as if Luttrell has died, the image and sound of his heart on a monitor secures for the viewer the assumed fact (given the film's title) that he will, indeed, survive. The film's tagline "never out of the fight" is reiterated here by Luttrell in voice-over. Luttrell's restored pulse transitions into the instrumental introduction of Peter Gabriel's cover of Davie Bowie's "Heroes," which includes a heartbeat-like double pulse in pizzicato low strings that transfers the effects-derived sound of the heart monitor into the film's score. To more character-driven ends, *American Sniper's* (2014) very minimal score activates a subtle representation of a heartbeat to suggest the ethical struggle Chris Kyle faces when confronted with killing women and children. The buildup to his first kill—a woman and her son approaching US soldiers with a bomb—is shown twice, first at the film's opening (where his first shot launches an extended flashback to Kyle's childhood and adult years before going to Iraq), then later to mark that the film is now set in the present tense. As he prepares to shoot the young mother and her son, a low beat begins in the score. This kind of adrenaline-suggesting touch is very unusual—indeed, only one of Kyle's other encounters as a sniper includes it. Late in the film, during his fourth tour, he shoots a man about to launch a rocket-propelled grenade. A young boy steps out of the shadows and picks up the dead man's dropped weapon. Kyle whispers to himself and the boy, "Don't pick it up," while a heartbeat pulses low in the score. The boy drops the weapon and runs; Kyle releases; the score falls silent. Kyle's coach at the shooting range counsels him to pull the trigger in the space between heartbeats. Confrontations with the need to kill women and children make finding that space difficult—but not impossible—for Kyle. His first kill is a boy; his second, a woman. His spotter responds to the death of the latter with the words "fucking evil bitch," perhaps speaking for some in the audience. Kyle shuts him—and those who agree—down, saying, "Get the fuck off me."

Heartbeats are metered, rhythmically organized along a consistent, duple pattern. Heartbeat-like moments in PCF scores impose a meter that either communicates the subjective experience of a soldier or—more often, especially in twenty-first-century PCFs (and action films generally)—administers a dose of adrenaline to the viewer. If the goal of military drum cadences historically was to keep marching men together in time, then the goal of much action-movie music in the twenty-first century is to set the viewer's heart beating and pulse racing. Such metered music—*beat-driven* proves a better term—goes apparently unheard by the characters in these films, who negotiate dangerous and active environments with cool aplomb. When applied to the narrative context of the PCF—where death matters—beat-driven scores create a variety of effects. But these effects, like the thrilling music heard on the helo insertion in *Zero Dark Thirty* (2012), are for the viewer. This music goes unheard by the people inside the world of the film. It is for us—those watching and listening, those along for the ride.

Black Hawk Down (2001) is the defining beat-driven score. Not only does the film feature almost continuous music, but the music is relentlessly beat driven. Outside a few veil cues and one extended melody—titled "Leave No Man Behind" (see the next chapter)—the score is an assemblage of beats reminiscent in their structure and timbres of various electronica music styles. Composer Hans Zimmer himself mentioned "techno music" as a source for the score. A four-on-the-floor beat pattern, a signature electronica and electronic dance music matrix, predominates no matter the provenance of the timbres heard in Zimmer's active textures. Like veil cues, *Black Hawk Down*'s many beats draw on a variety of distinct sounds: "exotic instruments" and "heavy rock guitar," to list two also named by Zimmer.[7] And while Zimmer suggested that these were sorted according to the two sides of the battle, the film generally moves too quickly between sides to keep such distinctions intact (as noted in chapter 7), creating an ambiguous context where, for example, Zimmer might link techno elements to the "high-tech American Rangers" while a film critic might hear instead "sinister Arab techno."[8] In the end, the goal seems to have been a continuity of beat-driven scoring, placed with prominence but not preeminence in the mix.

With continuity as the beat-driven score's steady state, any moment when the beat stops acquires tremendous power. The best example in *Black Hawk Down* comes when a soldier, in the midst of a firefight in an exposed position in an alleyway, looks to the ground and sees a severed hand. The beat stops, as if even the score registers surprise at the sight. The soldier picks up the hand and puts it in his pocket as the beat drops right back in: little to no time to contemplate, only time to act. *The Kingdom* (2007), like *Black Hawk Down* almost continuously scored, includes a stopping of the beat that engages the viewer early on. The film's opening credits are a brilliant audiovisual summary of the history of Saudi Arabia and the United States, a mini-documentary combining multiple informed narrators, some futzed to sound like radio or television voices. The mix of real and specially recorded narrator voices—Will Lyman from the PBS show *Frontline* is heard briefly—combine with documentary footage, informative titles arranged timeline-fashion, and visually realized statistics to offer a crash course in the history of oil as a driver of American foreign policy. The pace of the narration and the visuals is perhaps intentionally too fast. And the techno-style music by composer Danny Elfman, mixed on par with the voice-overs, adds yet another level of intensity. With occasional percussive hits on the titles, the music only makes it harder to follow the welter of information and images. Viewers expecting an exciting Jamie Foxx action flick might just give up trying to follow what is clearly historical background and wait for the story proper to start. But even if viewers just let it wash over them, the combination of voice-overs, visuals, and music communicate a confusing and dangerous history between the two nations, a relationship with tension at its heart. The chain of voice-overs falls completely silent just before the representation of September 11, 2001, an animation of a passenger jet flying into the Twin Towers, taking the screen to black. Elfman's driving beat stops completely on this blackout: sonically, the moment of impact is represented as a ceasing of sound. A more aestheticized representation of 9/11 as a history-stopping event with a long backstory is hard to find. The titles assume that no one in the audience needs an explanation for what's represented.

Zimmer's score for *Black Hawk Down* is built on the juxtaposition of different beats. Here the contrast between film score and score album is instructive. The score album for *Black Hawk Down* includes twelve tracks by Zimmer. Only one—"Hunger," first on the album—reflects at any length the score as heard in the film. The others are extended presentations of beat textures heard in much shorter form in the film. For example, "Synchrotone" lasts more than eight minutes on the album, whereas in the film this distinctive, sizzling beat is heard six times: initially for a minute and a half over the end of the mission briefing and the Somali informant marking the target building, and after this never for more than a minute, in once case for a mere ten seconds. A similar situation obtains for the funky yet colossal track "Tribal War" and the jangly and exotic track "Chant." These distinctive beats drop in and out of the film as needed. Most of Zimmer's beats are not connected in any clear way with individual characters or plot strands. On the score album, beats heard briefly across the film are offered up in rounded musical forms, from which a DJ might easily sample. Is the album a remix of materials from the score? Or is the score the result of sampling from the extended tracks on the album? And what about music on the album but not in the film? Are these beats Zimmer made then didn't use (or need) or that director Ridley Scott rejected? Zimmer, in his words, scored the film "just after Sept. 11" and tried not to let 9/11 "influence the way I was working."[9] The differences between film score and score album might reflect this intense and contracted working context, with the beats serving as charged sonic units easily dropped into place for seconds at a time.

Other PCFs with beat-driven scores suggest a more through-composed approach, with composers crafting long, layered cues—often heard without change on score albums—that track the shifting intensity of a given film's narration. John Powell's *United 93* (2006) and *Green Zone* (2010), Danny Elfman's *The Kingdom*, and Nathan Furst's *Act of Valor* fall into this category. A combat sequence from *Green Zone* is representative of these scores.[10]

Early in the film Chief Miller learns from the Iraqi civilian Freddy that a meeting of former Iraqi army officers is taking place nearby. Frustrated in his search for WMD, Miller says, "I wanna get something done" and Powell drops in a low, spare, triplet-pattern beat running at 121 bpm. This beat backs up Miller's instructions to his team, which are partially heard in voice-over on images of the team loading into vehicles commandeered for the mission. Miller's voice is completely calm. His men are mostly focused. When one asks how they know they're not headed into a "fucking ambush," Miller shuts him down: "We don't. Get your fucking game face on." Powell's score abruptly drops down as well. The triplet figure cuts out on a shift to a less regular, drier rhythm, a musical "game face" of sorts. Beat-driven scores, built on layers of sound, can respond with great economy to dialogue or action by removing or adding a layer.

The music continues, at lesser prominence in the mix, on a cut to the Iraqi officers' meeting. The beat stills entirely while the general, a major character in the film, speaks. Behind him, Powell's score waits at the ready on a low pedal tone that develops into ominous three-note gestures. (One of the few differences between this cue in the film and the track "Meeting Raid" on the score album is a shortening of this long pedal on the latter. In preparing music from the film for listening, Powell—who produced the album—perhaps recognized that the pedal held for the general's words in the film lasted too long in a music-only context.) After the general says, "Then we fight," a new, slower beat drops, a looser texture moving at 90 bpm. This texture is wiped away on a cut to Miller and his men approaching in two cars. The cars cross the screen in time with the beat, the sound of each passing vehicle folding into the score, facilitating the initiation of a new beat—a nervous eighth-note figure—moving at 180 bpm. Powell uses the tempo of the shot of the two cars to double the momentum of the beat, increasing the viewer's sense of the impending battle. Similar intimate connections between beat-driven scores and diegetic events—vehicles driving past, guns being fired—are common. The tempo and texture of the beat changes abruptly twice more on discrete points of action cued by Miller: he cries out the command "Go, go, go"; he kicks in the house's front gate. Powell doesn't always speed the tempo up. Shifts to a slower but thicker or more aggressive texture also work to increase tension. These shifts on points of action suggest a musical correlative to what SEAL Matt Bissonnette describes as "throttle control. We go fast when needed, but then go back to being slow and quiet."[11]

Once Miller and his men are inside the house, Powell peels off a few layers of the beat without slowing the tempo. The goal of the entire previous few minutes—bringing these two parties within shooting range—is nearing. The soundtrack quiets a bit, anticipating the gun battle to come. Then, a burst of drumming anticipates the initiation of fierce automatic weapons fire from Miller's team. Here, music precedes rather than responds to action. On the action, during ten seconds of gunfire, Powell's score falls silent, handing off to the effects track entirely. (On the score album, a huge, jangly, metallic beat fills these ten seconds. Powell perhaps scored the gunfire but director Paul Greengrass and the film's mixer chose to duck it down entirely, giving the action climax of the sequence to the guns alone.) When the firing stops and the chase continues, the score reenters at the fastest tempo yet—282 bpm—before ceasing abruptly and entirely when Miller watches the general escape in a car.

The close coordination between action and the beat described here obtains for almost all of *Green Zone*. This adrenalized music undergirds action in a step-by-step manner, finely shaped to the edit of the image track. The resulting music has a satisfying musical shape when listened to in isolation on the score CD, where Powell both adjusted its proportions and filled a sonic climax given to the effects track in the film with analogous musical material.

How does this adrenaline-fueled beat add meaning to *Green Zone*? Who is it for? Given the frequent cutting between the Americans and the Iraqi officers (and a group of their wives and children), it is difficult to locate it definitively in the hearts of anyone in the scene. The score tracks the Americans' actions and the general's words but not in a way that suggests that anyone in the film's world is feeling it. This music, like the sampled beats stitched into a whole by Zimmer in *Black Hawk Down*, is definitively for the viewer. Given their visual style, both films might have unfolded in a more documentary-like manner with no music at all. Both feature sufficiently dramatic and noisy action that music—especially busy beats mixed with prominence—would seem an unneeded distraction. Yet both films' soundtracks include a persistent musical layer, a beat-driven score fundamentally oriented toward the viewer. On a formal level these scores provide sonic continuity for disjointed crosscutting on the image track. In aesthetic, expressive, and ideological terms, these beat-driven scores elicit a heightened physical response in the viewer and generate excitement at—and forestall reflection on—combat action. The heartbeat in question is the viewer's.

But beat-driven scores need not serve as devices for quasi-physical immersion. They can also offer necessary distance from especially fraught on-screen events. *United 93*—like *Green Zone* a Greengrass picture scored by John Powell—utilizes beat-driven scoring to trigger reflection in a very specific action context. Powell's score is restrained in tempo, texture, and volume. With one exception—a bump in volume and musical energy in response to the line "Planes, plural" (admittedly on a cut to a shot of a jet engine)—Powell refrains from responding to specific dialogue or actions as he does in *Green Zone*. Indeed, there are few such actions to track in *United 93*'s narrative. Perhaps the most distinct musical moment comes when the score falls silent on a pulse of brass as the second plane hits the World Trade Center (as seen from an airport control tower in Newark and on television). There are no sound effects for this event, instead a general stilling of the score and soundtrack. As in the animated titles for *The Kingdom*, the opportunity to represent this particular September 11 sound was passed on by the makers of *United 93*. The silence after the image of the impact in both films represents the stunned reaction after the event rather than the sound of the event itself. In analogous fashion, the crash of United 93 in *United 93* is rendered as music, not effects, although without any image of impact. Again, comparison of film score and score album proves revealing.

The final soundtrack mix of *United 93* seems to have lowered or eliminated strong rhythmic elements originally in the score. The score CD includes many elements not audible in the film, where the score is usually kept low, creating a more subliminal disturbance for the viewer, perhaps more felt than heard given the dominance of low frequencies. The broad outline of events in *United 93* was surely well known to most of the film's original audiences. As summarized by the British novelist Martin Amis in 2008, "On United 93, the passengers were told about the new reality by their mobile phones, and they didn't linger long in the old paradigm. . . . No, they knew that they weren't on a commercial aircraft, not any longer; they were on a missile. So they rose up."[12] Greengrass's presentation of the passengers on their phones then rising up is decidedly restrained, as the discussion of the line "Let's roll" demonstrated. Powell's musical contribution to the film's final minutes is worth considering, as it is a beat-driven music that—like Alexandre Desplat's for the helo ride into and out of Pakistan in *Zero Dark Thirty*—is aimed squarely at the viewer's experience of the events. Only here, music, heard with increasing presence as diegetic sound slowly falls away, activates reflection on unfolding action with a known ending.

Once *United 93*'s secondary narrative in various air traffic control centers has been resolved with the closing of US airspace, the film remains solely on the doomed plane. The passengers, aware that three other hijacked planes have hit their targets, make final calls to their loved ones, pray (as do the hijackers), and plan their counterattack. Powell's music for this sequence is rhythmic and fast—about 190 bpm—but restrained and waiting, with long swells of brass that go nowhere. This music works as part of what Robert Burgoyne has called the "adrenalized stasis" of *United 93*.[13] Indeed, most of the music in the film and the experience of the film as a whole can be understood within the notion of stasis. There isn't much to be done but watch events unfold—events that are unknown to those in the story but known to the viewer. A beat-driven score, with its capacity to activate the viewer's body at a physical level, perfectly answers the need in *United 93* to present a series of events as both dramatic and known ahead of time, both unfolding in mystery and charged with excess meaning. The score consistently provides a subtly placed position both with the narrative's characters (hijackers included) and outside the narrative, where the historic dimensions of the day's events can be experienced at a respectful distance.

After almost ten minutes of calls, prayers, and planning, the score falls silent for a few seconds then restarts, at a slower tempo, with the passengers' counterattack. The action from here to the end is furious—the plane is crashing throughout—but the score remains generally calm. The only sonic expression of panic, beyond the screams and shouts of the passengers and hijackers, is a diegetic sound effect, itself strongly metrical: the cockpit alarm, a fast electronic pulse (also heard in *Black Hawk Down* when helos are crashing). In sharp contrast, the final cue—titled "The End" on the score album—unfolds at a steady pace. On the album a riot of percussion—chaotic fills and riffs, none setting any larger pattern—are heard against a slowly and regularly moving string line at the top of the texture. Powell did not try to catch specific events of the final struggle: there are too many and the sequence as cut is too chaotic. As mixed for the film, almost all of the percussion "stem" was turned down or off. The chaotic noise of the struggle stands in for the percussion lines and all that remains audible for most of the sequence is that top line, which rises and rises in pitch and prominence within the mix. It's an excruciating more-than-five-minute combat sequence with an end the audience knows, a violent end communicated sonically rather than visually. In the film's final shot, the camera—operating from an unattached POV—turns to look out the cockpit window toward the advancing fields of Pennsylvania. On a sudden blackout Powell's rising string line resolves downward by a fifth, a conclusive melodic motion further strengthened by the cessation of the beat. The resolving line—closing on a final C sharp—renders the sound of the plane hitting earth. Importantly, by the time the plane crashes the soundtrack mix has tilted toward near-complete diegetic silence.

Who is the music for in the final five minutes of *United 93*? The slow-moving, slowly rising string line with its low but present pulsing beat cannot be said to characterize the passengers or the hijackers. Both groups are engaged in a furious struggle at odds with the almost processional unfolding of Powell's top string line toward its conclusion. Instead, Powell's music—apparently adjusted by Greengrass in the mix—asserts a kind of rhythmic calm over the scene. Stephen Prince's conclusion about the film's depiction of the killing of the pilots and stewardess early on applies here as well: the music and mix can be understood to function as part of the filmmakers "trying to protect the feelings of the families of United 93's passengers and crew."[14] The film does not ask the viewer to get on United 93, but instead to observe what happened there. Beat-oriented music, especially at the close, offers a position outside the events from which to watch that is—importantly, in this PCF by an English writer-director—neither sentimental nor patriotic.

Beat-driven scoring also facilitates the representation of intelligence work as a form of combat. Many GWOT PCFs show soldiers and their kind peering at computer screens and video monitors. Subtle beats add sonic dynamism to the visually static image of a man or woman staring at a screen. Powell provided an early model in the cue "Miller Googles" from *Green Zone*. This thinking music runs at 113 bpm, with a big battery of percussion pounding away in what sounds like the next room. A very slow rising line and a slight increase in volume overall builds energy across an almost two-minute span. The line resolves at the close, when Miller (and the viewer) has absorbed the content of an Internet search. Similar thinking music is heard over a montage of the FBI team members in *The Kingdom* each doing slow, methodical forensic work.

A sequence of gradually more active thinking beats shapes the middle third of *Zero Dark Thirty*. The section begins when Osama bin Laden's courier Abu Ahmed al-Kuwaiti's name is first uttered, a moment sonically highlighted by an unseen helo and the entrance of restrained but still active beat-driven music in the score. Over the next seventy minutes, Maya searches for and finds the Kuwaiti—consistently called Abu Ahmed in the film—and follows him to what she correctly believes is bin Laden's compound in Abbottabad. The initial thinking beat accompanying and characterizing her search runs at 122 bpm. Immediately after Maya writes down the name Abu Ahmed, the beat enters very low in the mix and ties together a just over two-minute stretch of Maya watching multiple interrogations, most involving "enhanced" techniques. Here she comes to the conclusion that Abu Ahmed is close to bin Laden. The sequence and cue conclude on a cut to a toaster: a slice of toast pops up and Maya's hand reaches for it. This point of punctuation is analogous to the Japanese artillery piece fired at the close of the beach scene in *Flags of Our Fathers* discussed above. In both cases, an explosive event—one very small, one very large—marks a point of articulation and a shift of soundtrack mode after a musically organized stretch of film representing a particular aspect of a given war.

The 122-bpm thinking beat returns after Maya manages to get Abu Ahmed's mother's phone number. An analysis of patterns in who calls the number begins, including a montage of phone circuits and server farms, all underscored with the thinking beat, which is here tied to the word *tradecraft*—a chapter title flashed on-screen. In the GWOT, intelligence tradecraft is combat: Desplat's thinking beat characterizes such work as cinematic action. The beat simultaneously signifies the passage of a long stretch of time and, in its unpredictable small melodic motifs and dark, slightly sinister harmonic content, the painstaking work of intelligence gathering and analysis in a shadowy context. The beat returns some minutes later behind Maya's verbal summary of her conclusions for her colleagues (and, of course, for the viewer). Desplat's tradecraft beat is sufficiently interesting to function on its own but restrained enough to work behind important dialogue.

A new thinking beat enters the score after Abu Ahmed and his white SUV have been identified, his movements now tracked by a picket line of Pakistanis along the highway. A montage of their work—explained by Maya in a voice-over (which turns out to be the prelapped sound of her reading aloud an email summary of the search)—is accompanied by an especially layered tradecraft beat running at 172 bpm. As we get closer to bin Laden, the tempo and energy of Desplat's beats increase. Soon the film arrives—abruptly—at bin Laden's compound. *Zero Dark Thirty*'s long middle section prepares the viewer for the satisfying sight of the familiar (to many viewers) house in Abbottabad. (When I saw the film the first time, I spontaneously pointed toward the house, as if to tell Maya, "There it is. You found it.") The tradecraft beats drop out of the score at this moment, their work done.

Thinking beats grant a certain kind of GWOT combat needed cinematic urgency. As such, they are inherently pro military, even if, as in *Green Zone*, the diverse government entities fighting the GWOT are often seen working at cross-purposes. On a broader level, beat-driven PCF scores open the way toward the return of explicitly military music, easily evoked in the building of a beat texture around a snare drum cadence. Still, PCF scores that tilt toward traditional military marches are remarkably few. Predating electronica-referencing scores, the composer Nick Glennie-Smith briefly touches on a march cadence in *We Were Soldiers* (2002). Interpolation of Joe Strummer and the Mescaleros' "Minstrel Boy" for the end titles of *Black Hawk Down* puts a slow, snare-driven march into that film—a sound not heard in Zimmer's narrative score.

Only *Lone Survivor* goes all the way. The film's score, by the American post-rock band Explosions in the Sky, brings in a snare drum–heavy cadence on two occasions. The first ("Seal Credo / Landing" on the score album) plays behind the recitation of the "Ballad of the Frogman" and a montage of SEALs prepping for and heading into battle on a beautifully shot flotilla of five Chinooks (the kind of image that always indicates close Pentagon cooperation). Here the massive battery of snares suggests a recruiting commercial. The second snare cue ("Murphy's Ridge") accompanies Lieutenant Mike Murphy's sacrificial climb to an exposed position where his satellite phone will work, enabling him to finally call for help. After he's placed the call, and immediately after the surrounding Taliban forces begin shooting him—when it becomes apparent that he will die and that he is, in effect, already dead—a strong snare slips into the already-driving beat. This snare moves to the front of the mix on a complex series of shots linking all four SEALs, two of whom die at the same instant: Danny Dietz, shot at close range (we see the leader of the Taliban raising his rifle) and Murphy, shot from a distance, the impact of the bullets shown on his back and rendered in a quasi-musical sound effect as the final beat of the cue. Luttrell and Axelson, the other two SEALs, witness these events. Murphy's death recalls that of Elias in *Platoon*: both men are in an upright position, where the impact of the bullets that kill them can be seen with clarity, although the former faces away from the camera, while the latter moves toward it. The contrast between *Platoon* and *Lone Survivor* inheres in the films' very different plots leading to the witnessed death of a sympathetic character and the contrasting musical accompaniments for these parallel moments. Elias dies to Barber's *Adagio for Strings* (see next chapter).

Murphy dies to the sound of a traditional military cadence. *Lone Survivor's* narrative allows for snare-drum heroics that frame the act for what it is: one soldier's meaningful sacrifice for his comrades. Significantly, the snare does not energize his heroic climb or his call for help. Instead it lends a beat-driven, rather than elegiac, sacralization to Murphy's last moments.

MALICK AND ZIMMER'S "ATTACK ON THE BIVOUAC"

The musicologist K. J. Donnelly has noted, "Sonic continuity is often the foundation for visual discontinuity, emphasizing the semiotic inter-reliance but aesthetic divergence of the two tracks."[15] As shown above, beat-driven cues play this role in the often extremely discontinuous image tracks of twenty-first-century action films. As often is the case, *The Thin Red Line* offers an example where a PCF trope is pushed to a breaking point—in this case, a strongly metered (but not beat-driven) cue heard in a context of near diegetic silence during sustained combat action: a running battle with US soldiers taking a Japanese bivouac full of sick, malnourished, even crazed soldiers. As Donnelly notes elsewhere, "Sequences that remove most diegetic sound often allow a strong interaction between a loud piece of non-diegetic music and a succession of images, returning momentarily to the aesthetics of silent cinema."[16] Such is the attack on the bivouac in *The Thin Red Line.* Zimmer provided an extended musical cue—"Attack on the Bivouac," one of the longest on the film's cue sheet—that serves as a structuring element outside the unfolding of violent combat events shown from both the US and the Japanese perspectives. The sequence is the antithesis of the immersive all-effects battles discussed in chapter 6.

Musically "Attack on the Bivouac" begins as rhythm only: a ticking sound, difficult to place as a musical sound coming from an instrument, more akin to the sound effect of a clock. This ticking was heard once before in sustained fashion in the score when the soldiers initially began their move into the interior of the island. It's introduced very early in the film, on Witt's first voice-over, as a diegetic sound effect linked to the clock in his mother's death room. At the start of the bivouac cue, the ticking is thickened by soft, high wind and metal percussion sounds—again difficult to place as entirely acoustic instruments, and again a mix of effects and musical sounds. The tempo is quick: 240 bpm if the ticks are felt as the pulse; 120 bpm if they are understood to be eighth-notes against a quarter-note pulse (the latter interpretation makes more sense as the cue unfolds). The tempo of the cue is set here. It will not flag or rush for the next eight minutes, despite radical increases in the tempo of both action and cutting.

The ticking accompanies two of the film's most startling voice-overs, neither assigned with any clarity. The first, a conversation between two Americans:

"You seen many dead people?"

"Plenty. They're no different than dead dogs—once you get used to the idea. You're meat, kid."

The second seems to be assigned to a Japanese soldier who is almost completely buried: only his face appears above the soil, like a mask tossed on the ground glimpsed through smoke. The voice-over seems to be the voice of Elias Koteas (not in his role as Captain Staros), although this recognition comes only analytically, when listening to the film with eyes closed. A shot–reverse shot sequence suggests that Witt hears the Japanese soldier's voice, which asks, "Are you righteous? Kind? Does your confidence lie in this? Are you loved by all? Know that I was, too. Do you imagine your sufferings will be less because you loved goodness? Truth?"

As the voice-over ends, a low, slow-moving string line begins. This eleven-measure melody—spelled out in eleven whole notes—will cycle ten times, accompanying the entire battle sequence as an underlying ostinato (see musical example 1). The tune is odd. Its length is strange—most Western-style melodies are eight or perhaps twelve measures long. Its range is constricted—held within an interval of a fifth. Its harmonic implications are modal and therefore vague to the tonally conditioned ear: the second note (the "A") is the prime and the biggest leap is a third, denying any strong bass motion that a leap of a fourth or fifth would bring. A regular if hypnotic movement between adjacent or near-adjacent notes, cycling at an irregular phrase length, this musical underpinning is realized texturally as a grand passacaglia or chaconne—musical works that unfold over many repetitions of an unchanging bass line ostinato—initially mixed with the sounds of weapons, then in a context of diegetic silence and voice-over. Major transitions in the battle action are synced to the end or beginning of the tune. Zimmer's "Attack on the Bivouac" accompanies, or better fundamentally structures, a combat set piece moving at a furious pace with complicated, difficult-to-process shifts of camera angle and POV. Here the image track is organized at the level of the score by a musical structure of transparent simplicity. Given the almost aching beauty of the orchestration—mostly strings—the battle action is aestheticized in the interest of exploring both the American soldiers' and the enemy's response to the extremes of combat.

MUSICAL EXAMPLE 9.1.   Ostinato from "Attack on the Bivouac," *The Thin Red Line*, composed by Hans Zimmer

The shape of the battle aligns with the shape of Zimmer's grand ostinato-based cue: as detailed below, film form and musical form are intimately connected.

- Iteration 1 plays in a context of diegetic quiet, with only the sounds of men grabbing ammunition or loading their weapons.

- Iterations 2 and 3 play over American soldiers moving through a fog-filled jungle toward contact with the Japanese (shown prepared for battle at the end of iteration 1). The diegesis remains quiet: no jungle sounds, music dominates. This lyrical interlude is interrupted by the first bullet, which marks the transition between iterations 2 and 3. A soldier hears, indeed seems to watch, the bullet fly past him; the crack of its firing goes unrepresented. Partway through iteration 3, the firing begins with a tremendous crash of guns and very rapid cutting of hand-to-hand combat, including bayoneting. The score does not respond to this increase in the tempo of the action and cutting except to slightly grow in terms of texture and volume.

- At the start of iteration 4, Malick begins to frame the battle in running POV shots that follow individual American soldiers through the bivouac, a space that will never be clearly delineated but remains a dynamic zone of movement. These men run and fire; the camera runs after them, keeping them in the center of the frame, their actions providing a point of reference within the surrounding furious action. The identity of the men we follow is only vaguely suggested, especially for a first-time viewer. Also included is a brief POV from behind a Japanese soldier manning a machine gun in a trench.

- At exactly the start of iteration 5, in the process of following yet another American soldier through the melee, the camera swings right, distracted by the sight of a sick Japanese soldier being defended by another, also evidently ill soldier weakly wielding a bayonet. Both are screaming, though their screams are not in the mix. In an odd abandonment of the camera's previous following of a soldier, here the camera turns away from active combat and toward the plight of the suffering pair. At exactly this moment, a countermelody of three stepwise descending quarter notes enters on the soundtrack. The countermelody, with its suggestion of a sigh figure, is filled with consoling pathos and suggests the elegiac register. Cutting the sequence in sync with the iterations of the eleven-note ostinato allows the score to respond to new emphases in the images at a moment that is meaningful in musical terms.

- Over iterations 5, 6, and 7, the dire situation of the Japanese becomes the central focus. The start of iteration 6 marks a moment of disturbing engagement with a surrendering Japanese soldier. He looks directly at the viewer, backing away as the camera—from the POV of an unspecified armed American—advances. The viewer is put in the position of a US soldier, afraid in the context of battle but also confronted by an enemy who is helpless. This is not the battle the Americans—or likely the viewer—expected. The Japanese are malnourished, naked, bordering on the loss of reason. Individual Japanese are laughing, crying, provoking individual Americans to just shoot them. One is meditating. But there are still threats to be dealt with: snipers in trees, soldiers defending trenches. The US soldiers are still fighting. The Japanese are still in the process of giving up. The battle ends with the suicide death of a Japanese soldier by grenade. It detonates—filling the screen with a flash and a brief image of the soldier's face—on the eleventh and final note of iteration 7, a moment of musical arrival that clears the soundtrack for a return of the ticking heard at the start.

The disjunction between music and image across iterations 5 through 7 puts the viewer in an odd position. We are, at once, trying to make sense of the scene—like the American soldiers, whose point of view dominates the camerawork and cutting—and called upon by the music and sound mixing and the evident distress of the losing Japanese to contemplate the sad state of the enemy, who present vignettes of liminal humanity. The contrasting tempos of the very fast image track and the measured and moderate score create a tension for the viewer that defies easy assimilation, even on repeated and out-of-context viewing.

- The suicide-by-grenade explosion restores the ticking from the start and the ostinato tune briefly drops out: the death of the Japanese soldier is the only conclusive event in the narration of the scene. But the tune reenters on its fifth note as a kind of coda to the battle. Iterations 8 through 10 are heard in a context of diegetic silence. The action remains violent and evidently noisy: close images of Japanese screaming, Americans shouting, all without sound. By now the melody, having underpinned many minutes of screen time, is subliminally familiar. Over this distanced image track, the unassigned Southern voice-over comes in asking his characteristic philosophical questions: "This great evil—where's it come from? How'd it steal into the world? What seed, what root, did it grow from? Who's doin' this? Who's killin' us? Robbin' us of light and life. Mockin' us with the sight of what we might've known?" Iteration 10 concludes on the image of Witt lifting up a Japanese soldier: a soldiers' pieta blessed by the confirmatory final three notes of the ostinato. This image of succor and mercy is interrupted by a close-range gunshot—a diegetic blast intruding on the score, which has begun the tune again only to fall away into sustained chords. Only here does the metered music of the battle come to an end, on a seamless transition into an interpolated classical music cue: Charles Ives's concert work *The Unanswered Question* (a moment taken up in the following chapter).

The attack on the bivouac uses slow-moving, deliberately metered music to accompany furious combat action. Much of the previous hour of *The Thin Red Line* builds to this moment of confrontation. This is ostensibly a war movie: this should be the battle we've been waiting for. But when they take the Japanese camp, the Americans discover an enemy destroyed by hunger, ready to die by their own hands. This representation of the Japanese contrasts mightily with stereotypes found in almost all combat films. In striking fashion, Zimmer gives elegiac pathos to both the American soldiers—none of whom are presented as gung-ho warriors, most of whom look terrified themselves as they take the camp—and the Japanese soldiers—a defeated company, many mourning the loss of their friends. Zimmer's long, ostinato-based "Attack on the Bivouac," to which Malick aligned key moments of screen action, effectively shapes the mentally and morally challenging experience of this unusual metered combat sequence from start to finish.

10. Elegies

"Genre is alive," says Jeanine Basinger.[1] So, while, as Elisabeth Bronfen notes, "We can only partake of war [at the movies] by adapting the real event to the pathos formulas and codes of genre cinema," it is also the case, as Dana Polan writes, that "we might grasp the evolution of the combat film as the history of a genre in which experiments in form and in content are both at work, pushing individual works to say new things about the experience of war and to find new forms in which to say those things."[2] Several GWOT PCFs experiment with framing intelligence gathering as combat content. The World War II PCF cycle formally juxtaposes aged veterans and young actors. And with implications for all combat films to follow, the Vietnam War forced Hollywood filmmakers to come up with new "pathos formulas" to express the (new) national experience of defeat. *Platoon* (1986) is the most important "individual work" of the last forty years to find a way to "say new things about the experience of war." *Platoon*'s preeminent place in the history of the PCF comes with its invention of the elegiac register, an innovation in both form and content. This chapter traces the expressive path of the elegiac register across the full history of the PCF and beyond the cinema into the commercial public sphere.

PLATOON AND THE INVENTION OF
THE ELEGIAC REGISTER

Platoon begins in darkness and in silence. The silence breaks first. Samuel Barber's *Adagio for Strings* sneaks in on the title card "An Oliver Stone Film," slowly growing in volume, filling the soundtrack at the mordant Old Testament epigraph: "Rejoice, O young man in thy youth." As *Platoon*'s first image fades in—a C-130 transport plane slowly moving toward the viewer—the sounds of a busy airfield come up as well, competing with the Barber, which continues to unfold its strings-only, contrapuntal texture, never drowned out by the sound effects, which plunge the viewer into a dangerously noisy setting.

A loud squeak accompanies a cut to the plane's rear hatch. The ramp lowers slowly to reveal a group of fresh-faced cherries—"new boys," Stone calls them in the script—who come blinking into the sunshine.[3] The first sight they see is a cart loaded with full body bags. A soldier—Gardner, the first to die—asks, "Is that what I think it is?" Some watching *Platoon* might be asking the same question. This first line of dialogue positions the viewer as one of the "new boys," whose ground-level experience of the Vietnam War forms the center of *Platoon*'s narrative.

A sergeant appears, directing the new arrivals to follow him and offering the greeting, "Okay cheesedicks, welcome to the Nam." As the "new boys" fall in line, a group of "veterans" pass by, moving toward the C-130, departing the war after having survived their yearlong deployment. *Platoon*'s main character, Chris—one of the "new boys," played by Charlie Sheen—exchanges looks with one of the veterans in a dynamic crosscutting sequence. As called for in the script, the veteran's eyes are "starved, hollow, sunken deep in his face, black and dangerous."[4] He also seems to be smiling, full of knowledge Chris can't begin to imagine. Here at the start, *Platoon* hails such veterans as also part of its audience.

Barber's *Adagio* continues across a "hardcut" (Stone's word) from the airfield to the lush green jungles of Vietnam viewed from a literal helicopter shot. The viewer is riding into the jungle battlefield. The smooth chop of the rotor blends better with the still-playing strings than did the jarring noises of the airfield. The screenplay prescribes on the hard cut, "A rushing SOUND now. Of frightening intensity, an effect combining the blast of an airplane with the roar of a lion."[5] Had this plan been followed, *Platoon* would have begun with a jolt. Instead the jungle is first seen as both beautiful and melancholy and—importantly—of the past. The moviegoer is given no sense of how Chris might be feeling but is instead exhorted by Barber's *Adagio* to look upon the lost war's battleground with a kind of stoic recognition, a jungle mise-en-scène for American failure that, alas, must be revisited. Thomas Larson writes, "Within minutes, the movie is memorializing the men, and the music seems to know it."[6] In fact, the music is the expressive agent doing the memorializing.

Another hard cut drops the viewer to the jungle floor, where the camera looks up through tall trees toward the sun. Stone describes the shot: "Rays of morning light peeking through the cathedral dome of the jungle."[7]

(Similar shots in *The Thin Red Line* [1998] and *The Pacific* [2010] consecrate tropical rainforests as quasi-wondrous places where American men have been sent to die in the midst of terrible beauty.) All is still. The helicopter is gone: the soundtrack quieter and the still-playing, vaguely religious, strings-only music clearer for its absence. The opening cue ends, as it began, in a dominant position on the soundtrack. Barber's *Adagio* fades out on the end of a phrase—at a musically meaningful point—and the quiet hum of insects comes up on the effects track. The credits resume. We are in the jungle, slogging through mud and underbrush and insects, passing decaying enemy dead, on the first of the four patrols that structure the film's narrative. (Stone's abrupt transition from helo to jungle floor recalls Michael Herr's words in *Dispatches*: "Flying over jungle was almost pure pleasure, doing it on foot was nearly all pain.")[8] *Platoon*'s first few minutes land the viewer in Vietnam and deposit him (or her) on the battlefield, the entire journey conducted under the sustained, serious, elegiac musical escort of Barber's *Adagio*.

Composed in 1936 as the slow, middle movement of Barber's String Quartet in B Minor, the *Adagio for Strings* had long been an iconic piece of classical music when Stone used it in *Platoon*. The *Adagio* gained an independent existence and launched the composer's national career in 1938, when Barber arranged the work for string orchestra for a radio broadcast led by the conductor Arturo Toscanini, under the title *Adagio for Strings*, op. 11. The seven-minute piece quickly became an audience favorite, often played at times of national crisis. It was heard widely on the radio after the deaths of Franklin D. Roosevelt and John F. Kennedy. Jacqueline Kennedy programmed the *Adagio* as part of a broadcast memorial concert to her husband. In his book about the *Adagio* titled *The Saddest Music Ever Written*, Thomas Larson documents use of the work in concerts after the explosion of the space shuttle *Challenger* in 1986 and at a service for the victims of the bombing of the Alfred P. Murrah Federal Building in Oklahoma City in 1995. Larson details American conductor Leonard Slatkin's "agonizing ten minutes and twenty seconds" version at the Royal Albert Hall on September 15, 2001, a performance simulcast across the United Kingdom live and preserved on YouTube in two videos for a combined nine million-plus views as of fall 2015.[9] Writing in 1966, Eric Salzman pinpointed the work's success with audiences this way: "It has the curious distinction of becoming not merely a 'classic' but entering into general musical awareness in an almost subliminal way: it is one of those universally accepted and recognized sound images that are identifiable by people who have no concept of what it actually is."[10] The descriptor *elegiac* attached itself to Barber's *Adagio* as early as the 1970s, when Royal S. Brown described the work as "deeply elegiac."[11] The Barber scholar Barbara Heyman similarly calls the *Adagio* "elegiac" in her entry on the composer in *The New Grove Dictionary of Music and Musicians*.[12] The primary definition of *elegy* in the Oxford English Dictionary reads, "A song of lamentation, *esp.* a funeral song or lament for the dead." The critic Nicolas Slonimsky also connected the *Adagio* to funeral music: "The dirge-like serenity of the *Adagio* suggests tranquility in grief. Unexpectedly, this unpretentious movement from a string quartet became a threnody . . . selected again and again to be performed on occasions of public mourning."[13] This music—the antithesis of a military march—is spotted eight times in *Platoon*, and similar music resounds in combat films up to the present. This sad but beautiful, mournful music is the sound of war movies after Vietnam.

Stone's use of the *Adagio* in *Platoon* would install the work as a musical touchstone for a memorable (and easily spoofed) war movie moment: the slow-motion death of Sergeant Elias, played by Willem Dafoe, who dies with his arms flung high in the air, evoking Christ on the cross. Elias's death is a highly stylized and musicalized sequence. The mix balances the sounds of helicopters with Barber's music, as at the film's opening. Stone slows down the action and turns up the Barber as Elias reaches to the sky, then crumples to the ground. Slow motion lets the viewer absorb the violence of Elias's death, the bullets' perforation of his soft flesh. The Barber frames the moment as sacramental rather than narrative. He will not be rescued; in his dying he is already dead. Elias's slow-motion death takes a relatively long time: Barber's music opens up this temporal extension which, if the images are analyzed closely, shows Elias raising his arms and falling to his death several times. Elias's definitive death is signaled by a bump in the volume of the helicopter noises as they pass over his body one last time.

As K.J. Donnelly has noted, "In terms of music and image, anything potentially goes with anything. Synchronization . . . has less to do with maintaining the illusion of cinematic representation than it has with forming a different aesthetic whole. This is most clearly illustrated in sequences that remove most diegetic sound to allow a pure and strong interaction between a loud piece of non-diegetic music and a succession of images."[14] Elias's highly aestheticized death—observed differently by the soldiers in the film (in a helicopter) and the viewers of the film (on the ground in front of Elias, practically speaking in the line of fire)—uses elegiac music to frame the graphic representation of the dying soldier's body in a manner that encourages reflection in the midst of action. The men in the helicopter witness Elias's death from above. The music communicates the intensity of their experience of this sight. The viewer watches Elias die at close range with a string orchestra drowning out the noise of the battlefield—the first of many violent yet lyrical deaths of an American soldier set to elegiac music. As John M. Kinder notes, "Wounds are more than metaphors. Modern war is about many things, but its most defining feature is the rupturing, wounding, and destroying of human bodies."[15] An enduring connection between elegiac music and graphic images of the broken soldier's body was forged at this moment—one reason it has remained iconic.

As Elias falls and the helicopter leaves the scene, Barber's music arrives at a series of four intense, high, fortissimo chords, heard several times in the film. Here, these chords score an exchange of looks between Chris and Sergeant Barnes, who shot Elias in cold blood and told Chris he was dead. Chris turns his head to look at Barnes exactly on a change of chord. Preexisting musical form offers a scaffolding for character confrontation. All four chords are allowed to play out—the musical structure of Barber's *Adagio* is honored—before the sound of the helicopter rotors, which we know in reality to be deafeningly loud, overwhelms the strings and brings the scene to an end. With Elias's death, Stone demonstrates how surprisingly nimble elegiac music can be when its place in the mix is aggressively, indeed quite noticeably, manipulated.

With only seventeen cues in the entire film—not counting the trio of popular music interpolations discussed in chapter 4—Barber challenges Georges Delerue, the film's credited composer, for dominance of *Platoon*'s score. As shown in chapter 8, Delerue's eerie, tense veil music proved effective and influential. But Delerue also composed an *Adagio*-esque main title and several dramatic cues for *Platoon*, using the temp score—Barber, of course—as a model.[16] In trying to capture the essence of Barber's music, Delerue comes perilously close to simply copying it.[17] Stone used three of Delerue's faux Barber cues behind voice-overs spoken by Chris early in the film. Delerue's music matches the Barber's deliberate tempo, strings-only orchestration, and stepwise melodic motion but lacks the distinctive contrapuntal texture. Stone surely understood the danger of overusing the *Adagio* and the need to employ less intensely expressive music in the film's exposition. Indeed, after the opening credits the *Adagio* goes unheard for fifty minutes, almost half the film. Delerue's functional Barber-like music serves its purpose early on.

But once *Platoon*'s characters and central conflict have been established, Stone sets Delerue's copycat cues aside. Barber's *Adagio* returns at the film's narrative, thematic, and structural tipping point, the sequence of atrocities perpetrated by American soldiers at a Vietnamese village, reminiscent for many of the My Lai massacre of March 1968. Chris's intervention in an attempted rape during this scene was discussed in chapter 2 as an example of ABA form. Vietnam memoirist Philip Caputo describes participating in the burning of a village, the scenario for which maps onto the scene in *Platoon*: "It had been a catharsis, a purging of months of fear, frustration, and tension. We had relieved our own pain by inflicting it on others. But that sense of relief was inextricably mingled with guilt and shame. Being men again, we again felt human emotions." Later he notes, "Strangest of all had been that sensation of watching myself in a movie." But while Caputo experiences the "delirium of violence" in combat by way of films like *Sands of Iwo Jima* (1949) and *Battle Cry* (1955), he writes that there was no movie equivalent in his memory for the sense of shame and amazement he felt at doing what he did at the village.[18] Stone's use of Barber to close out the village atrocity answered a cultural need for veterans, like Caputo and Stone, to re-experience the war in a realm of myth-making cultural discourse—at the movies—in a representation where innovative music and mixing could open a space for reflection in the midst of action representing the troubling specifics of the American experience in Vietnam.[19]

Chris's voice-overs after the village scene are accompanied by Barber, almost always in a manner that begins and ends on a rest or breath mark in the *Adagio*. (On one occasion, the *Adagio* is interrupted mid-phrase by gunfire: sound effects, signaling engagement with the enemy, call a halt to the reflective music). The film's end titles, which are elided with Chris's final voice-over, use an almost complete playing of the work.

A grunt-level film telling a tale of military and moral defeat, *Platoon* demanded a new sort of movie music. Stone reached for an elegiac instrumental piece from the American classical music canon: Barber's *Adagio for Strings*. Barber's *Adagio* structures the narrative, ascribes meaning to important events, provides the viewer a refuge from the violent world of the diegesis, alters the status of the characters from active subjects to figures for the viewer's contemplation, and sets a tone of elegiac reflection at the film's start and finish. As used in *Platoon*, Barber's *Adagio* proved an effective, if easily copied, accompaniment to a story of American failure, a failure Christian Appy describes as "not just failure to achieve the war's stated objectives, but failure to preserve the broad conviction that America was an exceptional force for good in the world."[20] Hollywood had little experience with such a narrative and no generic musical codes on hand to express it. Interpolating a piece of serious, classical music worked well, and also elevated the film's action-movie content, bumping *Platoon*—and the combat genre—into the prestige category where critical praise and attention at the Oscars might be won. (It's worth noting here that, according to film composer Basil Poledouris, Stone initially planned to score *Platoon* with "The Battle Hymn of the Republic." Poledouris suggested that Stone use something "sad," specifically the *Adagio*, instead.)[21]

DEFINING THE ELEGIAC REGISTER MUSICALLY

Barber's *Adagio*, used in multiple films after *Platoon*, was effectively off-limits for subsequent combat movies.[22] But Barber's music and Stone's use of it proves central to the history of the PCF. Elegiac music following *Platoon's* example—original music composed on the model of Barber's *Adagio*, deployed within the narrative following Stone's practice—has played a substantial role in the form and content of the PCF and, in larger terms, in the representation of the American soldier in Hollywood feature films and audiovisual public culture. The use of musical form to shape film form described in *Platoon* recurs again and again in subsequent entries in the genre. When elegy begins, the image track is often shaped to follow the score.

Before examining uses of the elegiac register in post-*Platoon* PCFs, the musical form and content of elegiac movie music must be more fully described. As given above, the Oxford English Dictionary's principal definition of *elegy* includes two uses of the word *song*, suggesting how easily the elegy as a memorializing literary genre might be transferred to music. Further OED descriptors include *mournful, melancholy, plaintive,* and *pensive,* words that point toward the larger cultural work of the elegy as involving reflection of a particular kind: remembrance that is thoughtful, touching, tinged with sadness, and occupied with the reality of ultimate loss, the fact of death. Understood as a literal funeral song, elegy serves personal and corporate functions: moving the listener individually and internally while also marking the act of shared remembrance that is a funeral or memorial service. The elegiac register in war movie music transfers these functions into the perhaps unlikely space of the movie theater, turning the multiplex into a place of mourning.

As with the veil and beat-driven registers detailed earlier, the elegiac register can be defined musically as a flexible set of compositional choices drawn upon by composers and perhaps requested by musically astute directors. These include matters of tempo, orchestration, texture, harmonic style, and melody.

1. Tempo

The elegiac register employs moderate to slow tempos, maintaining a reserved, stately, deliberate, steady quality of motion.

2. Orchestration

A primary timbral marker of the elegiac register is the use of string instruments only. This distinctive consort sound—limiting the sonic palette to instruments in the same family—contrasts with the more varied timbral array of the full symphony orchestra or the military band, the historical norms for Hollywood war-movie scores. Innovation in the elegiac register has often occurred at the level of orchestration, adding hybrid strains to the register that depart from the founding example of Barber's strings-only *Adagio*.

The important hymn-elegy hybrid lends a more traditional heroic cast to the register by adding instruments associated with the military, such as brass, drums, and choral voices. This hybrid originated in the mid-1990s, as the elegiac register developed in the Vietnam films of the 1980s was being applied to other American wars. In his final cue and end titles for *Courage Under Fire* (1996), the composer James Horner supplemented strings with restrained brass (more horns than trumpets) and drums (timpani, not snares). Similarly orchestrated elegies followed in *Saving Private Ryan* (1998), *We Were Soldiers* (2002), and *The Pacific* (2010). The hymn-elegy hybrid, usually employing the major mode, delivers the sober reserve of the elegiac register in an affirmative spirit that explicitly supports military values without, however, suggesting a march off to war. All four PCFs with prominent hymn-elegies can be read as celebrating traditional concepts of military valor, even as they represent modern combat as an equivocal zone where chaos and chance rule and individual soldiers are broken. Its easy applicability to the World War II cycle is evident.

A second hybrid style involves the use of synthesized string and string-like sounds rather than actual stringed instruments. Heard in several twenty-first-century combat film scores, this synth-string hybrid registers the impact of digital music making on Hollywood practice and the decline of the hyper-classical, orchestra-centered score dominant from the late 1970s through the 1990s. It also lends a distinctive sound to films about America's wars in the Middle East.

3. Texture

Elegies are composed of layers of independent voices that play off one another in contrapuntal fashion. This is not to imply that elegies are not melodically oriented. Combat elegies typically have a top or prominent voice that can be heard as the primary voice. Indeed, a readily recognizable melody or tune is paramount, allowing the composer to run expressive changes on a musical theme the score has lodged in the listener's ear. The difference is that these tunes are supported and surrounded by other musical lines that play against the perceived leading voice, creating an intricately wrought texture that suggests a musical seriousness that gets transferred to the action on-screen. Strings-only instrumentation enhances the sense of several lines moving along independent paths.

The contrapuntal texture of the elegiac register tilts many film elegies toward art or concert music. Only classically trained composers are likely to have the knowledge to write independent lines that follow the rules of voice leading and treatment of dissonance. Elegies by composers without such training are often less complex in terms of their actual counterpoint but instead simulate the style, sometimes by the addition of an accompanying voice that has its own melodic integrity. Virtually all of Hans Zimmer's themes in *The Thin Red Line* employ this kind of simple counterpoint.

4. Harmonic treatment

The elegiac register is generally harmonious. Consonance rather than dissonance is emphasized. Simple triads dominate. Chromatic harmonies, jazz chords, and tone clusters are absent. In this respect, elegiac combat movie music sometimes suggests early eighteenth-century tonal harmony, where the dissonance of two voices pushing against each other in a carefully prepared manner is enough to generate the pleasures of discord giving way to concord. The enriched harmonic language of the nineteenth century, with its sinuous chromatic lines and diminished-seventh chords, synonymous with tension or suspense in Hollywood music since the silent era, is set aside in the elegiac register, which reaches for an in-context "purer," less hackneyed harmonic idiom that, to the moviegoer's ear, sounds fresh.

5. Aspects of melody

Elegiac melodies usually move in careful stepwise motion, often incorporating descending stepwise figures that tap into historic associations between such so-called sigh motives and the expression of sorrow. In harmonic terms, sigh figures and stepwise motion in a contrapuntal texture encourage the kind of rigorously prepared dissonances typical of pre-nineteenth-century harmonic practice. Sequences (short melodic figures repeated at different pitch levels in an ascending or descending pattern) and walking bass lines are common features as well: John Williams's "Hymn to the Fallen," for example, has both.

Most elegiac melodies are completely diatonic, with composers limiting themselves to pitches in the chosen scale or "in the key," an expressive restraint that generates melodies that do not reach for distant or surprising effects.

At the front end of elegiac melodies, the slow-moving nature of the register is highlighted by two kinds of sustained openings. Suspended melodic openings begin with a held note in an upper voice—initially heard alone—to which a bass note is added. The bass voice renders the initial upper voice dissonant, necessitating melodic motion—usually by step—that in turn gets the music moving. A second, widely used strategy involves preceding the start of an elegiac melody with a pedal tone, usually in the bass. Such cues can initially sound like the veil cues discussed in chapter 8. The difference comes with the entrance of multiple, clearly melodic voices, which activate the contrapuntal texture typical of the elegiac register.

Elegiac melodies almost always come to clear points of rest. While individual phrases may feel quite extended, due in part to the characteristic slow tempos, elegies are usually organized into rounded phrases that reveal a predictable and balanced structure after just one hearing. Many elegiac themes employ clear cadence formulas that signal that the melody is coming to a close. Several such formulas have historical roots in the eighteenth-century styles elegiac counterpoint often evokes.

The elegiac register takes one of two musical forms: wholly elegiac cues or elegiac episodes within more varied stretches of music. Elegiac cues and episodes alike have the potential to shape film form in pronounced ways, framing events as they unfold within the narrative and directing the viewer toward particular interpretations of images, sound effects, and dialogue (which often takes the form of voice-over). The next section details how the elegiac register opens stretches of fundamentally musical time dedicated to somber reflection, part of the PCF's larger memorializing function. A subsequent section shows how elegiac music has the power to redirect the perceived tone or mood of entire films, both early on and very late in their narrative unfolding. Examples from *Courage Under Fire* and *Green Zone* (2010) show the potential of the elegiac register to challenge the viewer to think critically about combat and US policy, demonstrating the expressive muscle of this music. The final section explores how the register's formal capability—a kind of expressive pause on the action—has been used to frame representations of the broken male body, a visceral trope of the subgenre that literally pictures the act of patriotic sacrifice.

ELEGIAC CUES AND EPISODES

Elegiac cues are continuous pieces of music that are sustained in tone with unambiguous beginnings and endings. With their clearly articulated, balanced phrases and prominent cadence formulas, elegiac cues have the capacity to forcefully shape a given stretch of film by virtue of their formal musical characteristics. Three examples follow, each demonstrating a different combination of elegiac music, effects, and dialogue. The makeup of the mix plays an important role in the power of the elegiac register to shape a given reflective moment.

One of the many elegiac cues in *We Were Soldiers* accompanies the journalist Joe Galloway taking photos on the battlefield. The battle is at its most intense after a napalm canister is dropped by an American plane within the American lines, grossly burning Jimmy, a friendly Asian American soldier who earlier tells Galloway his wife is giving birth that very day. Jimmy's wounds are displayed in extremely graphic fashion—the flesh on his legs liquefies as Galloway picks him up; his face is half destroyed, yet Jimmy keeps on talking about his wife and child. After helping Jimmy onto a medevac helicopter, the shaken Galloway takes up his camera and begins to capture what he sees. In a music video–like montage that takes the film completely outside the (literal) heat of the battle and into a zone of diegetic silence, images of Galloway taking photos are superimposed over the black-and-white still images he is in the process of capturing. Coming in over a string pedal, an "ethnic" flute plays a descending, mournful tune—one of two elegiac melodies heard across composer Nick Glennie-Smith's score—with an independent string line below creating a contrapuntal texture. The tune's four balanced phrases are played through, after which the entire tune is repeated, the flute replaced by a solo on the erhu (heard without processing or distortion) supported by a fuller complement of strings. Suspensions add harmonious tension to the music, which moves with a slow dignity. Galloway's two-minute photo shoot stands outside the narrative for its highly stylized visual language, its clear musical form, and diegetic silence in the mix. The viewer is invited to briefly put aside the immersive, graphic violence of *We Were Soldiers* and take a visual journey through more objectified horrors of war to the strains of aching, beautiful, pathetic, harmonious elegiac music. The elegiac cue sets off the sequence from the narrative, in the process interrupting the flow of the film considerably.

Ennio Morricone's score for *Casualties of War* (1989) includes cues titled "Elegy for Brown" and "Elegy for a Dead Cherry." Each uses the same long-breathed melody, and each is tied to the fatal wounding of an American soldier—the film's two close-range US casualties. "Elegy for a Dead Cherry" comes late in *Casualties of War*, accompanying a speech by Eriksson. Looking on the body of a "cherry" who steps into a booby trap—grotesquely pierced by pongee sticks—Eriksson reflects on the moral demands of the situation for American soldiers. It is difficult to concentrate on his words, which closely paraphrase the actual soldier quoted in Daniel Lang's article on which the film is based, while also attending to Morricone's finely made cue.[23] The graphic sight of the cherry's body, shot from above as a spectacle of a vicious war, also competes for the viewer's attention in a scene overloaded with visual, verbal, and musical content. Similarly, "Elegy for Brown" plays during continuously unfolding action that interferes with possible apprehension of the scene by way of the music. Brown's resistance to dying—he shouts over and over, "I'm a armor-plated motherfucker"—trumps the music, unlike the scene of Elias's death in *Platoon*, where elegiac music fills the soundtrack and the cries of the dying man are seen but not heard. Brown's elegy plays on past Brown's exit and into the return of the mundane concerns of the war, as the remaining men in the squad are given new orders. The finely wrought end of Morricone's elegiac cue mismatches this thrown-away dialogue by characters unconnected to the intense sequence still drawing to a close in the score.

Morricone's two "Elegy" cues are similarly scored for strings and horn, with the horn serving as an element of color rather than a solo voice. Both cues begin with a soft and low bass pedal, followed by a high, sustained note that responds to a third independent, melodic voice. Repeated sigh figures mark the upper voice's leading line throughout. The upper and lower voices frequently move in stepwise contrary motion, pulling against each other, creating a con-trapuntal texture that sets these two cues apart from the modally inflected themes related to Oanh. The musical subtleties of Morricone's "Elegy" cues are best appreciated out of context, on the score album—where they are paired together as the final two tracks on the disc—or on several of the many discs of Morricone's movie themes arranged for concert performance.[24] In the film, director Brian De Palma gives the viewer too much sonic content, with dia-logue and elegiac music competing on the soundtrack to the diminishment of the latter.

The only even faintly melodic cue in Thomas Newman's score for *Jarhead* (2005) evokes the elegiac register by way of a stepwise rising melodic line in a synth-string sonority. The cue "Jarhead for Life" begins with a guitar vamp embellishing a B-major chord—a pedal figure that con-tinues throughout. A melodic line, sometimes with an accompanying line shadowing it a consonant third below, rises from F sharp (the dominant, a rather consonant starting point) to the tonic (B) an octave and a half higher. The melody is mostly stepwise, especially early on—in essence, a scale. The patterned decoration of the chord generates recurring moments of simple dissonance when single jarring notes in the ascending melody rub against the tonic triad. The harmonic tension created between the unchanging strummed chord and the rising line exploits the most basic sorts of disso-nance in the Western tonal system. The sense of arrival and completion at the end of the ascent is palpable, and the resonant texture rings with the comfort of a consonance found and sustained in this otherwise spiky, sour film.

The action accompanying "Jarhead for Life" is violent and body-centered. Corporal Alan Troy is being branded on the leg with a makeshift iron reading "USMC." Early in the film, Swofford is attacked by the men in the company, who hold him down and pretend they are going to brand his leg (a scene counterscored with the pop tune "Don't Worry, Be Happy"). The Marines do not brand Swofford, saying that he has to "earn it"—a phrase echoing back to *Saving Private Ryan*'s climactic words, "Earn this," spoken by the dying Captain Miller. Troy's actual branding occurs shortly after the company learns he is being kicked out of the Marines for not revealing a prior criminal record, a banishment understood by all to be a gross injustice, for the only place Troy has ever felt at home is in the Corps. The branding—accompanied by the elegiac cue "Jarhead for Life"—marks Troy as a true Marine in the eyes of his comrades. Swofford whispers in Troy's ear "You earned it," turning the phrase, taken from *Saving Private Ryan*, into *Jarhead*'s refrain. Soldiering as an identity that marks the body is here literalized in the brand and musicalized by way of the elegiac register in a twenty-first-century guise. Gradually, as the noisy action unfolds, all diegetic sound fades out, leaving just the rising musical line. Swofford's voice-over comes in to interpret the scene in a classic PCF mix recalling Chris's voice-over admiration for his buddies in *Platoon*. Except where Chris calls his platoon mates "the best of the best," Swofford takes a more intellectual approach, analyzing why the Marines believe themselves safe within their homosocial "circus" and ending with the sarcastic line, "But we are insane to believe this." *Jarhead*'s constant undercutting of the Marine ethos comes even here, on the heels of the score's only gesture toward elegiac wholeness.

The cues from *We Were Soldiers, Casualties of War,* and *Jarhead* analyzed above demonstrate the considerable expressive fungibility of the elegiac register, in both its musical content and its narrative functions. All three also point toward the centrality of the soldier's body as a visual motif with the power to activate elegiac music. This emphasis on the sacrificial soldier's body runs across the PCF, defining the genre's consistent concern with representing the embodied experience of modern combat and the actual bodily sacrifice of American soldiers. Elegiac music opens a space where such bodies can be looked upon and thought about.

Elegiac episodes are passages of elegiac music within longer cues encompassing several scoring styles. Since the typically slow-growing elegiac register must be activated within an ongoing musical texture, episodes test the ability of composers to move efficiently into a reflective musical mood, sometimes in the midst of continuous action. Two examples from beat-driven PCF scores follow.

In *Black Hawk Down* (2001), Zimmer makes room for the elegiac register by layering it into the almost continuous, driving musical texture of the score. In several sequences, a diatonic, strings-only, stepwise, slow-unfolding melody brings elegy into the mix. Recourse to the elegiac mode in this manner underwrites the serious intent of *Black Hawk Down*, marking it off from generic action-adventure movies, directing the viewer to take the cinematic representation of battle deaths—most reenacting actual American deaths—with appropriate seriousness. Stripped-down elegy makes room for contemplation of the sacrifices demanded by combat even as those very sacrifices are unfolding before the characters in the film and the viewer. The score memorializes the American dead even as the men fighting beside them do not have the time to stop.

Zimmer effects this running memorialization through the use of a theme heard in an extended six-minute version on the score album under the title "Leave No Man Behind." Two complementary melodies are heard successively at the start of the track. The first melody is the more active of the two, with four phrases of equal length, each using the same rhythmic pattern. Stepwise motion, a plaintive double-sigh figure in the third phrase, and a general sense of motion that goes nowhere—it's one of many futile elegiac tunes—projects a directly expressed, almost folk-like sadness that will not be consoled. This melody, in its second use in the film, accompanies the first combat death in *Black Hawk Down*. The tune begins in a solo horn as the Humvee convoy moves out after having loaded the prisoners who are the goal of the mission. A bell marks the downbeats and the tune feels, for a moment, like action music. But at the end of the tune, Sergeant Dominick Pilla, manning the .50 cal turret gun, is shot at close range by a Somali man on the street. Blood from Pilla's wound splatters on the camera lens—a pro-filmic PCF trope dating to mud splashed on the lens in the opening credits of *Platoon*. Pilla falls into the vehicle, his descent marked by slow motion and a sound effect that suggests a last breath leaving his body. The viewer knows Pilla has been killed: the melody and his life simultaneously come to a close. The melody restarts—in a slightly different orchestration with less horn, more strings—as Pilla's death is confirmed within the vehicle and reported on the net and up the chain of command. Close-ups of soldiers responding to the news follow as the tune plays out, ending on General Garrison back at the TOC just as the melody comes to a close. With the second restart of the tune, the bell marking the downbeats returns on a decisive sound-matching cut that puts the film back into action mode. "Someone get on that 50," cries the driver. Hoot climbs up. As he blasts away at the Somalis in the street, the melody plays a third time, coming to a close and yielding to a beat associated with the RPG-armed Somali who successfully shoots down the first Black Hawk seconds later. *Black Hawk Down* moves relentlessly on after Pilla's death. With no time to stop and mourn the first American death, Zimmer manages to slow the flow down a bit, structuring the sequence around three iterations of the "Leave No Man Behind" melody. Director Ridley Scott, of course, had to cut the film to match the tune. The orderliness of the tune—its equal-length phrases, its modal sound, its use of sustained strings—puts an elegiac frame around Pilla's death and, for the duration of its three playings, effectively shapes the unfolding of *Black Hawk Down*.

In *Act of Valor* (2012), composer Nathan Furst slips easily in and out of the elegiac register in a score otherwise dominated by tense veil cues and beat-driven, orchestral action music. The first "act of valor" involves the rescue of a kidnapped female CIA agent being tortured at a terrorist camp. Veil cues accompany the SEALs' stealthy approach. Upon their entrance into the building where the agent is being held, the score cuts out for some tense silence. Gunfire from the kidnappers initiates a driving action beat. As happens so often, the sound of gunshots—crucial battlefield signals—initiates a change in the score. When the SEALs have secured the building, the beat cuts out but the music continues, as Furst smoothly shifts to elegiac strings, enhanced by a wordless, wailing vocal. The shift in musical register matches a shift in action and a visual focus toward the gravely wounded soldier's body: the Chief removes the agent's bonds—her arms bound, her palms pierced—and tells her, "I'm bringing you home."

The beat begins again as the extract sequence starts—a truck chase where action music plays only when the kidnappers are in active pursuit. During this fast-paced battle on twisting jungle roads, the SEALs use an RPG to successfully and spectacularly destroy a pursuing truck. Furst pauses the beat for the drawn-out explosion of the truck. In the midst of this thrilling action, the wounded SEAL Mikey wakes up in the truck after having passed out upon being shot in the head. Mikey's disorientation adds further confusion to an already chaotic scene. Abruptly, all diegetic sound drops out as an elegiac episode replaces a beat-driven texture in the score. Mikey screams in agony—without a sound and in slow motion. The Chief, in command, looks on, registering the confused scene inside and outside the truck. Bit by bit diegetic sounds reenter: first, muffled sounds of machine guns firing at the pursuing truck; then, labored breathing; the music swells;

diegetic sound reenters in full; and Mikey is successfully calmed. The beat drops back in as the chase proceeds.

The elegiac episodes for the kidnapped agent and for Mikey—each associated with a wounded body—anticipate a full-fledged elegiac cue closing the jungle rescue "act of valor," a musical coda that opens a space for reflection on and with the SEALs. Having boarded the rescue boats, the sounds of battle fade and elegiac strings flood the soundtrack, in a cue titled "The Calm After." All diegetic sound fades out—in particular, the very noisy motors on the boats. In this space of diegetic silence and musical plentitude, the SEAL team members exchange long, meaningful glances, calmly contemplating the successful outcome of the mission.

Act of Valor demonstrates the capacity of the elegiac register to shift narrative focus from unfolding action to focused reflection on the fragility of the human body. The music escorts the viewer back and forth between the scene of action and a space for contemplation. Such shifts assume a narrative structure that anticipates that elegiac music will be added at the last, when the musical score is laid over the edited image track.

ELEGY, GENERIC REDIRECTION, AND APPEALS TO CRITICAL THINKING

Elegiac music has been used to redirect the apparent genre of entire films—moving them from action-adventure flicks into PCFs by musical means. In the process, viewers are enjoined to recalibrate their reception: to think critically about the military (*Courage Under Fire*) or about US military policy (*Green Zone*). Such interpretive turns—activated in large part by the score—show the elegiac register doing heavy work.

Courage Under Fire begins as an action movie centered on fighting men and their war toys—in this case tanks. But a tragic battlefield mistake during the first sequence—an instance of friendly fire perpetrated by Lieutenant Colonel Serling, played by Denzel Washington, the presumptive hero of the film—signals that it will not be an unproblematic tale, as fans of Washington's earlier work might have assumed given the film's title. What begins as a gung-ho flick abruptly turns toward the PCF subgenre. And even though the Pentagon refused to cooperate with the production on the grounds that the mutiny at the center of the plot would be "astonishing behavior for the all-volunteer, post-Vietnam Army," *Courage Under Fire* garnered important support from veterans.[25] The press kit included a letter from the president of the Congressional Medal of Honor society, who saw the film as a needed lesson for the times:

Courage Under Fire is a unique movie. It challenges our sensitivities and stimulates our minds without gratuitous violence and sex. . . . As a war movie, and *Courage Under Fire* is much more than a war movie, *Courage Under Fire* accurately teaches us of the intensity, confusion, chaos, fear, pressure and speed of modern combat. . . . *Courage Under Fire* is a must for all the leaders of our society, military and civilian, young and old. It is a must for all of us. It is a film which if heard and observed, not just listened to and seen, will prompt each of us to pause and take stock of our lives and the society in which we live.[26]

A radical change of tone in James Horner's score early in the film signals this "prompt" to "pause" on the soundtrack.

High-octane action music accompanies the opening tank battle, a thrilling piece of fully scored action filmmaking that Serling kicks off by leading his men in a prayer for protection followed by the injunction, "Let's kill 'em all"—an oft-repeated sentiment in the subgenre. After Serling mistakenly destroys one of his own company's tanks, commanded by his good friend Boylar, Horner's dramatic action scoring shifts to haunting strings-only cues, some with wordless female voices sighing in sorrow. This, and not action music, is the default sound of the score. Elegiac strings, sometimes with plaintive piano, accompany Serling's combat flashbacks and struggles with alcohol abuse (cascading strings a half hour into the film provide an especially intense elegiac moment). Not until late in the film, when Serling listens to a tape recording of his own battlefield response just after the incident of fratricide, does action music return—very briefly—to the score. The tape reveals how in the heat of the battle, moments after having mistakenly killed Boylar, Serling creatively determined a way to prevent further friendly casualties. Able to distinguish enemy from friendly in the lines opposite his position, Serling commands his crew to destroy an Iraqi tank. The successful hit receives a triumphant stinger that inserts just a touch of action-movie music back into the score. This does not, however, turn *Courage Under Fire* definitively toward action music. Just after the celebratory stinger, the film's long, slow, final movement begins, reinforcing the musical identity of *Courage Under Fire* as a PCF (see next chapter).

Two PCFs from the GWOT cycle use elegiac cues late in their narratives for purposes of generic redirection. In *Green Zone* and *The Kingdom* (2007), exceptional elegiac cues set within beat-driven scores call into question US prosecution of the "war on terror," asking viewers not to find resolution—impossible, in any case, with the conflicts still raging—but instead to reflect, however briefly, on the limited efficacy of American power and the underlying motives for US action against twenty-first-century enemies. Elegies near or at the ends of these films suggest a bid by the filmmakers to encourage viewers to think back upon the narratives through a particular sort of lens—one that raises fundamental moral questions about US actions in the region. In these films, the elegiac register leads not toward an emotional response to US soldiers but rather plants the seed of an intellectual response to US policy. *Green Zone* is discussed below; discussion of *The Kingdom* follows in the next chapter.

As noted earlier, *Green Zone's* continuous action narrative is accompanied throughout by composer John Powell's beat-driven score. Powell helped define the idiom with his scores for all four films centered on the character Jason Bourne, which, like *Green Zone*, star Matt Damon and were—in three of the four cases—directed by Paul Greengrass. (The *Bourne* scores lack elegies.) At the climax of *Green Zone's* action-packed third act—when Chief Miller finally corners the Iraqi general he believes knows the truth about the missing WMDs—the score dips briefly into an elegiac cue that presents the tragedy of the American misadventure in Iraq as a fundamental lack of intercultural understanding, a product of the misguided belief that the priorities of the US can be easily transferred to foreign others as just so much common sense. For at the moment when Miller has captured the general, Miller's interpreter, Freddy, shoots the general dead at point-blank range. The musical cue "Wtf" is named after Miller's verbal response to Freddy's action. "What the fuck did you do?" asks Miller, who is deeply surprised, even hurt, that Freddy, heretofore a sympathetic but not uncomplicated figure, would cut off his quest for the truth about the Bush administration's argument for the invasion. Freddy answers with a line that resonates not only back over *Green Zone*, but outward over the US audience watching. Freddy says to Miller, "It is not for you to decide what happens here," and the elegiac cue begins.

"Wtf"—with its string-synth sonority and melodic content that unfolds painfully slowly in a deliberate, suggestively contrapuntal manner—stands out markedly from the rest of the score. This music responds to Freddy's line. Miller may be the putative hero—and Damon the reason *Green Zone* got made—but as Freddy points out, witheringly for the American viewer, even a "can-do" action hero ultimately has no control in Iraq circa 2003. Greengrass pulls the rug out from under the film by rejecting the facile notion that the United States can manage events in Iraq, where even apparently sympathetic Iraqi partners act—to Miller's naive surprise—in their own interests. The music Powell uses to underline the moment gives the inherent dignity of the elegiac register to a foreign "other," a rare event in PCF history.

After Freddy departs, Greengrass punctuates the conclusion of the scene, and the now-failed quest at the center of *Green Zone*, in dramatic fashion. The camera pulls up and away, rising to a panoramic CGI view of Baghdad in flames—an especially bleak moment in post-9/11 cinema. Powell's score matches Greengrass's crane shot by rising in pitch, growing in volume, and adding ominous low brass. The hints of elegiac pathos bestowed on Freddy expand in a melodramatic fashion, shifting to Miller—and the United States—stuck in place. The cue ends not with elegiac comfort but with monstrous music for a battle still raging.

"SOMEBODY, GIVE US BACK OUR BODIES!"

Michael Hammond has noted, "The depiction of violence in contemporary Hollywood . . . is part of an aesthetic history in which the war film has often been involved," specifically "the drive to depict the physical impact of bullets, shrapnel and explosions on the human body."[27] When images of a soldier's body—usually wounded or dead—have been presented for the viewer's contemplation in the PCF, elegiac music has consistently been in attendance. Cinematic offerings of the soldier's body as a sacrifice for the nation fill the subgenre, from its inception to the present, going some way toward addressing the real Ron Kovic's cry, voiced in his 1976 book *Born on the Fourth of July*, "Somebody, give us back our bodies!"[28] In *Men's Cinema: Masculinity and Mise-en-Scène in Hollywood* (2013), Stella Bruzzi argues that "one of the many enduring pleasures of violent action movies is that, after the initial assault, they stop surprising us, and being inured to brutality is part of the satisfaction."[29] (Tellingly, Bruzzi mentions only one PCF in her book: *The Deer Hunter* [1978].) Two-thirds of the way through a semester-long course on the PCF, after watching about twelve films, one of my students remarked that the death of an American soldier on-screen had ceased to move him. The PCF cannot afford such numbness in its intended audience. Elegiac music helps activate emotional engagement, hopefully forestalling such impassive responses. Below I consider scenes from four films where the literal or figurative bodies of the dead are viewed by characters in the film and, of course, also by the film audience, who watch both the dead and those who look upon them. (The audience's double position of listening and watching listening soldiers listen, parsed in chapter 5, is repeated here in a different context.) These moments from *Saving Private Ryan, In the Valley of Elah* (2007), *The Thin Red Line*, and *Flags of Our Fathers* (2006) demonstrate the expressive leverage supplied by elegiac music, which has the power to point the viewer to larger themes surrounding sacrifice on the battlefield and the personal dimension of the death of a young man. A further two scenes from *The Thin Red Line* and *The Hurt Locker* (2008) deal with even more intimate moments when American soldiers touch the bodies of the dead. Here, too, elegy attends.

Only forty of *Saving Private Ryan's* 162 minutes include non-diegetic music, in all about a quarter of the film's run time.[30] Composer John Williams's generally restrained score sticks to moderately paced orchestral music mixed well to the back. Much of the score—including the important, self-contained end-titles cue "Hymn to the Fallen"—taps into the hymn-elegy hybrid first heard in Horner's score for *Courage Under Fire*. But in a few passages, including the introduction of the narrative score's main theme, Williams reaches for a strings-only, contrapuntal texture that falls squarely within the Barber-esque elegiac register. These moments picture or evoke the bodies of the fallen—whether specific characters in the story or the larger throng of American dead on the battlefields of Europe in World War II.

The twenty-three-minute, unscored, relentlessly noisy D-Day landing—*Saving Private Ryan's* signature episode—falls between two closely scored sequences set in official, orderly, quiet spaces with almost no dialogue. The first—its cue titled "Revisiting Normandy"—opens the film and forms one half of the present-day frame around *Saving Private Ryan's* historical narrative. The second—to the cue "Omaha Beach"—takes the narrative from Omaha Beach just after the D-Day landings to the war department in Washington, DC, where secretaries are typing letters of condolence to the families of the dead. This bureaucratic scene introduces the film's central narrative: the search for Private James Ryan, the last surviving son of a family with four brothers in the military. Williams slips into a strings-only elegiac episode once in both cues. Both episodes are directly connected to the visual topic of soldiers' bodies presented literally or metaphorically.

The strings-only episode in the cemetery sequence begins when the old Ryan takes his first step toward the field of tombstones. Horns and snare drum rolls come in on Spielberg's rising crane shot, which emphasizes the sheer number of the dead, but strings alone play while Ryan wanders into the midst of these markers. Descending violin lines cross with ascending cello lines in a dense contrapuntal weave that does not put forth a melody so much as mark the moment by textural means. The episode, unfolding in diegetic silence, is short, ending when Ryan falls to his knees. His family—and clarinets in the score—come to his aid.

The post-landing "Omaha Beach" cue begins with a preparatory pedal tone in the bass—a signal elegy may be coming—entering on the line "quite a view," as Captain Miller and Sergeant Horvath look out over

Omaha Beach. Spielberg cuts from an extreme close-up of Miller's eyes to a view of the beach Miller could not possibly be having: close-range images of American dead at the shoreline, their bodies washed by a tide tinted red with their blood. This eyeline match doesn't scan. The audience is, from here, detached from Miller's perspective, moved into a zone where personal rather than plot-centered interpretation of the images is encouraged. On the cut to the bodies, Williams introduces the primary melodic material of his score: first heard in the cellos, the "Ryan" theme is moderate in tempo, expansive yet rounded in its shape (two four-bar phrases with a two-bar tag that elides with the restart of the melody), and diatonic in pitch content with wide leaps and expressive sigh figures. An accompanying contrapuntal voice, also in the cellos, lends gravity to the whole. The melody is heard twice in its entirety over images of dead American bodies in the surf, lending dignity to the sight, at last giving the viewer time to contemplate the battle experienced in the film's sustained, immersive representation of the Omaha Beach landings.

As the elegiac main theme continues to unfold, Spielberg cuts to a different perspective, suspending the viewer above the iron cross stanchions on the beach, floating over the dead, zeroing in on one body sprawled on the sand apart from the others, the name "RYAN. S" stenciled on its pack. Williams directs the eye to this body by way of the ear with an added high strings pedal as the camera begins to move toward it. This body is the first clue to *Saving Private Ryan*'s narrative. But having finally hinted at the story to come—almost thirty minutes into the movie—Spielberg pulls back yet again, keeping the film's focus on the collective loss of life the landings entailed. The seen-from-above shot of Ryan cuts to an extreme close-up of an unidentified woman, who looks down as if looking at Ryan. She is, in fact, contemplating the American dead as part of her job typing condolence letters—another use of a not-quite-matching eyeline match. As Spielberg builds the mise-en-scéne of the War Department—working from close-ups outward to master shots—Williams's score takes a further turn toward the elegiac.

Leaving the "Ryan" theme, Williams modulates gently into a brighter key area as an intensified strings-only texture emphasizing the higher voices of the violins takes over. This mesh of violins complements a verbal counterpoint of male voices heard in a collective, layered voice-over—fragmented voices of officers expressing the bravery and contributions of the dead give the viewer access to the content of the letters being typed (as discussed in chapter 5). This novel adaptation of the ubiquitous PCF voice-over expresses a collective and official, yet still personal, grief that momentarily interrupts what had been the first concrete suggestion of *Saving*

Private Ryan's plot. The emergence of the Ryan narrative in the typing bay—a secretary discovers three letters to the same mother—unfolds slowly and visually, only after a more general sense of the loss of life on D-Day is expressed in sonic terms by way of the mutual counterpoint of officers' and elegiac string voices.

The body of Mike, the soldier and son at the center of *In the Valley of Elah*, is hacked to pieces by the Army buddies who murder him, burned in an open pit, then gnawed upon and scattered around an isolated field by wild animals. Mike's murder is not represented, but the film dwells repeatedly on Mike's remains, picturing his bones and burned flesh as discovered at the crime scene, as framed by police photos, as analyzed by a dispassionate coroner while Mike's father looks on, and, finally, as laid out on a tray to be viewed through a window by Mike's mother, "the emotional weathervane of the film," who insists on seeing her son's body.[31] She touches the window, says it must be cold in there where her son lies, and asks to go inside. (She is denied entry.) When a military investigator offers his condolences, she turns to him and says, "You don't have a child do you? Do you?" Her husband leads her away, down a long hallway. This and a stark scene where Mike's father picks up his son's body in a coffin on a dark loading dock are all the time the film allots to formal reflection on Mike's desecrated body. The score amplifies these moments, consecrating Mike's memory.

Composer Mark Isham surrounds images of Mike's remains with a special sort of elegiac music that stands out within *In the Valley of Elah*'s score as a whole. Most of Isham's score includes a low rhythmic level, a disturbing pulse troubling the mix. This beat is stilled during the strings-only music for Mike's body: a music that remains very soft, moves very slowly, stays in the mid to upper ranges, and communicates tremendous sorrow. The tentative and diffuse string voices, which move together in a solemn fashion, evoke gut-stringed gambas rather than the violin family. This elegiac music expresses the simple fact of Mike's loss—physically represented by his mutilated remains—rather than the mysterious circumstances under which he died. Particularly connected with Mike's mother, who is not included in the procedural plot, the meaning of this restrained example of the elegiac register can be summarized by the title Isham gave one elegiac cue: "A Family's Grief."

If *The Thin Red Line* has a narrative climax, the sacrificial death of Private Witt is arguably it. For this moment late in the long film, Zimmer provides serene elegiac music unheard to this point: a series of sustained, consonant chords, realized mostly in strings, moving slowly and harmoniously to a conclusive and drawn-out cadence that resolves on the image of

Charlie Company gathered around Witt's fresh grave. The music starts just before Witt is shot, as a Japanese soldier, rifle raised and ready to shoot, approaches him. The shooting of Witt is a slow-motion fragment—the sound of the shot that kills him rendered as a crashing wave—that immediately cuts to the image of Witt's body as a representation of his spirit: he swims in crystal blue water with native children while the harmonious chords roll on. Witt's body is never broken before the viewer's gaze. Instead, he escapes into a blue, liquid world. Witt's elegy closes with a drawn-out, satisfyingly complete resolution, recalling the score's first cadence point, which closed on Witt's first appearance. The contrast with the similarly beatified Elias in *Platoon* is telling. Witt and Elias alike embody a gentle, masculine care for others. But while Elias's gory, Christlike sacrifice is needless—the result of Barnes's treachery—Witt gives himself freely for his brothers and is last seen in a new, transfigured body. As he hoped early in the film, Witt meets death "with the same calm" his mother did. Both men, however, die before our eyes to elegiac strings in stylized soundscapes.

Clint Eastwood's *Flags of Our Fathers* theme is heard in both major and minor modes. During the film's key sequence, the minor-mode version stands in for a sight Eastwood spares the viewer: the desecrated body of a young soldier named Iggy. Many questions raised by writer Paul Haggis's fragmented narrative are answered during a series of lightning-fast flashbacks experienced by the flag raisers Doc Bradley and Ira Hayes during the Soldier Field war bond rally. The soundtrack cuts abruptly between John Philip Sousa's march "The Thunderer" being played at the rally and the noises of the battlefield as Bradley and Hayes remember how three among the men who raised the flag—Mike, Harlon, and Franklin—were cut down on a barren stretch of Iwo Jima. Repeated sonic crosscuts between Sousa's fatuous military march and the sounds of bullets ripping into human flesh cut to ribbons the platitudes of the film's war bond drive and simultaneously eviscerate any notion of bravery or courage on the modern battlefield. The three men die horrible deaths. Hayes and Bradley—the latter a medic—can do nothing but witness their friends' dying words and breaths. Bradley's final flashback reveals what happened to Iggy, a question raised in the film's first minutes and returned to in its last. Bradley flashes back to the nighttime battlefield where the film begins, still looking for Iggy. Another American takes him to a cave and shows him a dead soldier who turns out to be the missing Iggy. Bradley sees what the Japanese did to Iggy: the viewer only "hears" it as rendered by the main theme in the minor mode plaintively played on the piano with three mournful sigh figures in the horns lending a contrapuntal element. We look at Bradley looking at

Iggy for as long as the tune lasts. The sonic content of the Soldier Field flashbacks juxtaposes the jingoistic music of pre-Vietnam war films in its original form (the Sousa march), the visceral sound effects of the digital surround-sound era (which violently rip open the bodies of Mike, Harlon, and Franklin each in turn), and the elegiac register used as an aural stand-in for the soldier's body (which, in the case of Iggy, Eastwood chose not to represent).[32]

In two remarkable sequences, the elegiac register sounds when soldiers are confronted with dead bodies or body parts that must be touched. In both cases, the body belongs to a foreign other. These sequences from *The Thin Red Line* and *The Hurt Locker* use the elegiac register as an expressive lever to tilt the PCF into challenging territory for its American audience.

Two linked scenes involving a direct and disturbing physical confrontation between an American and a Japanese soldier receive elegiac emphasis by way of a dense, strings-only contrapuntal music not heard elsewhere in the score for *The Thin Red Line*. As described in the previous chapter, much of the furious and violent action during the attack on the bivouac unfolds with minimal diegetic sound; music plays prominently throughout, and the unassigned Southern voice-over articulates wonder at the scene near its close: "This great evil. Where's it come from? . . . Who's doing this?" Zimmer provides a lush musical counterscore to the chaos and violence of the sequence, which extends to both the content of the images and the editing. The viewer is not invited into the destruction of the bivouac but instead enjoined by the sustained music, the querying voice-over, and the non-continuity editing to contemplate and struggle to understand the scene, which unfolds horror upon horror.

Just after Witt is seen comforting a Japanese soldier, two figures take prime focus: a Japanese officer and an American soldier. The American, who briefly shatters the diegetic silence by firing his rifle point-blank into the back of a Japanese prisoner, is then seen guarding the officer, an older man who writhes on the ground in evident pain. The soldier, a powerful young man who stuffs his nostrils with pieces of a cigarette to cut the stench, holds a pair of pliers and toys with the officer. Like the music, the two men move slowly. They speak to each other, but neither understands the other's words. The American says calmly, "I'm gonna sink my teeth into your liver." The officer's words—among the only spoken Japanese in the film—go untranslated; his voice and visage are filled with desperate emotion. The American "speaks" a final time in voice-over (at least it sounds like him). We hear his thoughts—"What are you to me? Nothing."—as he turns, pliers in hand, toward an adjacent Japanese corpse, searching for gold fillings to add to his collection of such combat trophies. Just prior to the start of this slowly unfolding encounter, filmed from the perspective of a twisting Steadicam, the score seamlessly slides into Charles Ives's *The Unanswered Question*. The work is a kind of elder cousin to Barber's *Adagio*. Both feature strings moving in slow, consonant counterpoint. But the Ives moves *very* slowly and includes a second instrumental layer of solo trumpet calls, in a different key and time, utterly out of relationship with the strings, in a concert performance played from offstage. Two of the trumpet calls sound in the Ives as excerpted for *The Thin Red Line*, repurposed as perverse military signals marking the warped perspective of the American soldier.

Director Terrence Malick picks up this vignette some ten screen minutes later, and Zimmer provides striking elegiac music unheard elsewhere in the film. The American soldier sits alone—stripped to the waist, in the pouring rain—holding the bag of teeth he has collected from the enemy dead as if it were a sacramental offering. He trembles and shudders violently, then casts the bag into the dirt. Malick stays on the soldier for a long, long time, forcing the viewer to struggle to interpret the scene, to gain some meaning from the episode, as the soldier hugs himself in agony and looks to heaven. Zimmer's cascading sigh figures fall down upon the soldier—and on the viewer—as a comforting element during a liminal moment. The battlefield trophies he casts away mean something tremendous to this young man. Just prior to this long moment—made longer by elegiac music, which has no melodic shape and so continues with no hint of when it will cease—the Southern voice-over comments, "War don't ennoble men. Turns 'em into dogs. Poisons the soul." The relentlessly beautiful music offers a forgiving benediction of sorts for the soldier and gives the viewer distance from the scene, a remove not granted the American soldiers watching within the diegesis, who look upon their comrade in distress with impassive faces—unable, in context, to process what they see. Viewers of the film, assisted by the insistent, "meaning"-heavy music, are encouraged to react with greater engagement.

When Malick finally cuts away from the shaking soldier, the rain abruptly stops. Private Bell, the soldier for whom sensuous memories of his wife sustain his every move, walks along the airfield. In a voice-over addressed to his wife, Bell reflects on the challenge of staying human in the midst of combat: "You get something twisted out of your insides. . . . I wanna stay changeless for you." Zimmer's strings-only counterpoint washes over these words as it does the shuddering soldier in the rain. Bell and the nameless soldier share the same desire to resist the changes combat brings on those who face it. Zimmer's elegiac music doesn't cleanse these two so much as sonically connect and tint their stories, making their common struggle to stay human a noble one marked by music that responds to ugliness with elegiac beauty.

Zimmer's score, allied with Malick's lyrically shot and edited image track and the elliptical use of unassigned voice-over, supports a view of combat masculinity as a philosophical state, equal parts body and spirit and fundamentally touched by grief. Malick defines the soldier as a figure of deep emotion and existential pain. Malick and Zimmer build on Stone's use of Barber's *Adagio* in *Platoon* but take the elegiac register into more abstract, less historical territory. In *The Thin Red Line*, elegiac music defines the modern soldier not as an American man betrayed by his government and society—as in Vietnam—but as a prototypical modern figure, caught in an untenable situation where life and death struggle before his eyes, leaving him empty inside. Only the celestial Witt, his sacrificed body delivered whole and untouched into transparent waters, escapes to a better place.

Marco Beltrami and Buck Sanders's score for *The Hurt Locker* draws on the elegiac register during an especially intense dual-focus sequence: Sergeant James finding and removing a body bomb sewn into the corpse of a young Iraqi boy and the death by IED of the psychiatrist Cambridge who had been working with Specialist Eldridge, the youngest member of the EOD (explosive ordnance demolition) team. The two parallel events are linked by an elegiac melody heard four times across about five minutes, a rare lyrical touch in this score.

Deciding not to blow the body bomb in place as he had first planned, James instead removes the explosives from the chest cavity of the dead boy, who he believes to be Beckham, an Iraqi boy who sells DVDs at the base. James's distress is not expressed musically. His very evident pain—among the only times he shows any emotional vulnerability—goes unscored, including the moment when he tenderly closes the boy's eyes. At such moments we see, in director Kathryn Bigelow's words, "the attritional effects of war" on James: "So you see that he's actually not impervious."[33] A minor-mode melody made up of eight whole notes divided into two four-note phrases sneaks into the mix while James does the actual work of removing the bomb, a sequence that lingers relentlessly and graphically on the boy's body. Very low humming sounds are supplemented by a rhythmic pattern bringing a visceral heartbeat into the mix but not suggesting adrenaline. As James pulls the explosives out of the boy's body, the eight-note melody begins a second time, pulling against a pedal-like voice that suggests a contrapuntal texture, very unusual in this score that typically only throws out melodic fragments, if that. The mournful music continues as

James shrouds the boy's body in white—a nod toward Muslim practice—and carries him out of the building. James's decision to keep the boy's already-sacrificed body whole rather than blow it to pieces connects to the importance of children in narratives about the moral dilemmas of the overseas battlefield, a trope reaching back to World War II–era combat films and powerfully re-invoked in *Saving Private Ryan* (where an American soldier dies trying to help a French child) and many subsequent PCFs, such as *In the Valley of Elah*. The viewer is invited to watch exactly what James must do to the boy's body. That we must view a child's body—first desecrated by the bomb makers, then ripped apart by James—only adds to the intensity of the moment. Yet again, elegiac music comes to our aid, organizing a stretch of time when the bodies of the dead are put on display.

After a brief cut away to the psychiatrist Cambridge, a parallel scene discussed below, a third repetition of the eight-note melody plays while Specialist Eldridge and Sergeant Sanborn, the other members of the EOD team, watch and comment on James's exit from the building bearing the boy's body. This tune-shaped cue adds an independent bowed bass line, creating a genuinely contrapuntal texture more audible for the lack of hums or dissonant veils in the mix.

Intercut with the scenes of James and the boy is a tedious exchange between Cambridge, doing a ride-along outside the wire, and a group of Iraqis hauling rocks in a donkey cart. When Cambridge finally gets the group to leave—for their own safety, he says—he is killed by an IED left behind (and perhaps detonated) by the departed Iraqis. The squad observes the explosion from their Humvee: the original script called for Cambridge's head to be blown off and come flying against their windshield.[34] In the film, Cambridge's body just disappears, his helmet all that's left. Eldridge bolts from the Humvee. Sanborn and James follow, trying to prevent the distressed soldier from potentially setting off another IED. Eldridge screams, "I just saw him." James replies, "He's dead." On this dialogue, the eight-note melody associated to this point with the boy and the body bomb plays one last time, with yet more contrapuntal lines and set within a hushed mix of dissonant veils.

The above four-and-a half-minute sequence features all the musical strategies developed in PCF scores over the previous decades in complex combination: unsettling, dissonant veils; a rhythmic element suggesting a human heartbeat; a suggestively contrapuntal, melodically ordered texture evoking the elegiac register. The larger sequence is organized in musical terms by the rondo-like repetitions of the eight-note, mostly stepwise tune, which connects the boy's body—which James refuses to let be consumed by war—to Cambridge's body, which the IED vaporizes before the men's (and the viewer's) eyes. Sound designer Paul Ottosson spoke of the music in this sequence and the need to be careful in its use so as not to tip *The Hurt Locker* toward what he and Bigelow perceived as a too-manipulative beauty: "I loved that music. It has this very simple line in it and it's such a beautiful piece of music. At some point I discussed it with Kathryn because I was going to bridge the two cuts where he's carrying the boy with the music, but it became too much like we were trying to pour something on, that's because ultimate beauty is setting in."[35] Ottosson's concerns and reediting of the sequence can be understood as a desire to use the elegiac register while also holding off its power to anoint ugly images with "ultimate beauty," to keep the scene raw and difficult, to hold back the power of this especially potent movie music to give the viewer a way out or a step back from the film. The uncompromising nature of *The Hurt Locker* inheres in exactly such musical choices.

In the early years of the Iraq War, more than a few US corporations bought time on network television to express support for the troops. Among these public service announcements, one stands out for its condensed invocation of the elegiac register in a radically truncated time span. For the 2005 Super Bowl, Anheuser-Busch, makers of Budweiser beer, created a one-minute spot titled "Thank You." This short movie, usually available on YouTube, lacks any dialogue, directly invokes the deployment of elegiac music in PCFs, and echoes the opening moments of *Platoon* described at the beginning of this chapter.

The scene is an American airport terminal filled with middle-class travelers—a symbolically charged public space after 9/11. Six seconds into the spot, synth strings sneak into the soundtrack, which heretofore contains only muffled noises. About thirteen seconds in, someone among the passengers begins to clap. The first to applaud is unseen, although shortly after the sound begins a passenger rising to her feet to join in the applause is shown, anchoring the sound in the image. Only then does the viewer get a clear view of the soldiers, a multiracial group of men and women who pass through the terminal dressed in clean field uniforms, as ordered by the Bush administration for servicemen and -women traveling on domestic flights. The terminal has become a point of contact between the nation and its military, something like the PCF's transformation of American movie theaters into public spaces for the memorialization of wars past and present.

The elegiac music that signaled the soldiers' entrance begins to grow, as does the volume of respectful applause. At about the midpoint of the spot, individual soldiers—a black man, a Latino woman—are given passing close-ups. Their reserved, modest, appreciative responses to the applause can be registered by the viewer. On a cut to a small girl watching the soldiers pass—like the young Kovic watching veterans on parade in the opening of *Born on the Fourth of July* (1989)—the applause begins to rapidly duck out as the music grows in volume and texture. Soon music alone fills the soundtrack. Indeed, the entire second half of this one-minute movie features no diegetic sound. Synth-strings playing harmonious, slow-moving, vaguely contrapuntal music sound over the noisy image of American travelers clapping for the impromptu parade of soldiers. The music comes to a long, satisfying, drawn-out cadence on a title card reading "thank you," followed by a second card reading "Anheuser-Busch."

Platoon begins with "new boys" arriving at an airfield in Vietnam to the sound of violins. No one applauds their passage, which is attended by body bags and veterans headed home. The soldiers in "Thank You" are symbolically thanked in ways the veterans of Vietnam were not, but the music that plays as they move across the screen reaches back directly to Stone's innovative use of Barber's *Adagio* to tell a very different tale. As suggested by the gap between these two similar scenes, the elegiac register has proved an ideologically flexible approach to the scoring of war films, demonstrating the defining role this kind of movie music has played in the cinematic representation of the human cost of the nation's wars in the post-Vietnam era.

11. End Titles

Grief is a practical mechanism, and we only grieved those we knew.
—KEVIN POWERS, *The Yellow Birds* (2012)

He sat in his chair and thought of the dead.
—GEORGE PELECANOS, *The Double* (2013)

Music during the end titles can be heard as a narrowly formal, even industrial requirement of film form. Many PCFs transform this stretch of time into a space for sustained reflection on the core concerns of the subgenre: the soldier, the veteran, the proper response by the moviegoer as citizen to narratives of personal sacrifice for the nation. *Hymns for the Fallen* comes to its close with an examination of music for the end titles, both in its own right and as related to music at the moment of narrative closure. PCF directors and composers frequently use the ends of their films and scores to create a final, distinctly musical zone of meaning, carving out a space for listening in the cinema where audiences are invited to process the film through the medium of film music offered (in most cases) sans images.

Long end-title credit rolls are a rather recent cinematic phenomenon, arriving, by coincidence, concurrently with the PCF subgenre itself. In the studio era, credits were clustered at the start of a film, with ten or so title cards flashed on-screen to the accompaniment of music. This conventional opening created the need for composers to write "Main Title" cues: a stretch of music, usually less than two minutes in length and unconstrained by images (although sometimes marking the appearance of specific names on-screen). These mini overtures typically introduced the primary musical themes of the score and musically smoothed the viewer's passage into the narrative by means of a harmonically inconclusive ending. The very brief final credits of the studio era—a quick "The End" and sometimes an abbreviated cast list—left composers little room to be creative at the close. Extended opening titles and very short end titles remained the industry standard into the mid-1970s.

The shift to the current regime of long final credit rolls, virtually always with music, took place rather abruptly: an early example being the industry-transforming *Star Wars* (1977). Director George Lucas gave screen credits to individuals who would never have received acknowledgement in the studio era—for example, the makers of the film's special effects. The inclusion of all those names had aesthetic consequences for composer John Williams, who was given about four minutes of screen time to fill with music while the credits rolled. (This sort of musical cue was genuinely new. For example, Williams's score for *Jaws* [1975] does not have extended end-titles music.) Williams's cue for the close of *Star Wars*, titled "The Throne Room / End Titles" on the score album, provides a musical bridge from the end of the narrative to the credit roll. It begins with a fanfare announcing the ceremonial occasion that brings *Star Wars*'s heroes together one last time, a scene with no dialogue save R2-D2's last few bleeps and bloops and Chewbacca's final roar. Williams's music makes this somewhat awkward conclusion credible, even possible: it's hard to imagine another way to play the scene sonically. On a final group shot, the narrative world of *Star Wars* irises out on a massive musical cadence, launching an expansive playing, in succession, of two of the score's primary themes, the heroic theme for the brass associated with Luke Skywalker (heard at the top of the film) and the lyrical melody for the strings associated with Princess Leia. Williams's musical peroration on *Star Wars* rewards the viewer who keeps listening. Not only does it span the gap between narrative and end titles, formally confusing when exactly the film ends, but it also follows a balanced, musically satisfying plan. The "Throne Room" section acts as an extended fanfare for the end-titles section, which unfolds in an ABA (thematically Luke-Leia-Luke) form. The theme we want to hear again comes back in triumph when the credit roll reaches its end, at which point the music—and *Star Wars*—concludes at a still incredibly high volume. Clearly no one told Williams the only people listening at this point would be the cleaning crew.

Of course, many audience members were still in their seats at the end of the end. (I was.) Part of the allure of *Star Wars* on its original theatrical release was the film's use of Dolby stereo, which enhanced the sonic experience of the film—specifically the soundtrack's clarity, volume, and placement in the theater—and brought Williams's symphonic score into particular prominence. The thrill of listening to a symphony orchestra on a good sound system with the volume cranked up was part of the appeal of *Star Wars* from its opening seconds. This cinematic listening context for symphonic music—rather than hearing an actual orchestra live in a concert hall—has become the norm for most Americans in the decades since.

As Williams's "Throne Room / End Titles" cue suggests, in the era of long final credit rolls, the border between the narrative and the credits can be rendered ambiguous by the score, which has the potential to carry the expressive content and interest of a film beyond the close of the diegesis into a zone of experience that can only be described as musical. End-titles music invites audiences to close out their experience of a film by shifting into a listening mode, turning the cinema—a space for audiovisual experiences—into an auditorium—a place where music and listening are privileged.

Long final credit rolls literally make time within the span of a film for viewers to pause and reflect on what they have experienced, to think about the film to the sound of music. Music gets the last "word." Michel Chion calls this "the ritual of the closing credits ... a sort of airlock between the temporality of the film and that of daily life. ... We *need* this moment, even though it's narratively useless."[1] Chion celebrates the extended end credits "ritual" as a chance to hear sound unmoored from image in the space of the cinema. He does not, however, consider how the musical content of end-titles music might guide the viewer-now-listener in particular interpretive directions relative to both the film and daily life—the two realms connected by the "airlock" of the end titles. Different genres use this added space differently. The PCF has generally given over its end titles to a kind of post-narrative reflection during which viewers can "count the cost" of the story just seen. Music serves an important and flexible role in pointing the reflective listener toward a range of positions open for identification at the close of these films, which are, in essence, narrative meditations on the cost of citizenship as borne by the men who fight the nation's wars. In short, in the PCF the ritual of the end credits is far from "useless."

The musicologist Ben Winters offers helpful categories for thinking about music heard during the credits. In his 2010 article "The Non-Diegetic Fallacy: Film, Music, and Narrative Space," Winters coins the term "fictional music" to denote music heard during a film's narrative, when its construction of a diegetic world is up and running. His complementary category "extra-fictional music" applies to nonnarrative music, which Winters identifies as "perhaps only in credit sequences, where the cinematic frame and the constructedness of the fiction are openly acknowledged. ... In these situations ... the music clearly cannot be part of the narrative, though the transition between these states can be extraordinarily fluid."[2]

The transition from fictional to extra-fictional music, which necessarily marks the transition from narrative to end credits, happens in a limited number of ways. Many film narratives close with an unambiguous cadence in the score marking the moment when the diegetic world of the film goes to black. Such narrative cadences set a musical seal on both the film's fictional music and the film's construction of a coherent diegesis. Then, after a few seconds of silence—usually with the screen gone to black—the end titles and an autonomous end-titles cue begin. In a second option, the gap between narrative closure and the end titles can be spanned by a continuous piece of music, a musical bridge that serves as the only connecting element going across the divide between diegesis and credit roll. In such cases, a single music cue changes in status from fictional to extra-fictional in relation to the image track. Musical bridges fall into two types: those with internal cadence points marking the moment of narrative closure (such as *Star Wars*) and those without. Finally, the shift from narrative to credits can be marked in the score by two pieces of music seguing (or moving directly and without pause) one into the other. There is, of course, the option of closing out a film's narrative without music. This choice proves very rare in the PCF: only four end this way: *Hamburger Hill* (1987), *84 Charlie MoPic* (1989), *American Sniper* (2014), and all but the final episode of *Generation Kill* (2008). Once again the PCF is revealed to be a deeply musical subgenre, almost always using music at the moment of diegetic shutdown. It's even rarer to withhold music from the start of the end credits. Only *Generation Kill*, with its comms chatter as "music," takes this option, although the second half of the closing credits for *Redacted* (2007) and *American Sniper* scroll in absolute silence.

Winters's observation that the transition from fictional to extra-fictional music can be "extraordinarily fluid" applies to many PCFs. A common closing strategy across the subgenre involves the insertion of significant mediating elements between narrative closure and end credits, troubling the border between essential story content and the formal content of the end credits, often with the goal of engaging the viewer with larger questions beyond the fate of the characters or the resolution of the plot narrowly speaking. Mediating elements include: dedications hailing specific audiences, epigraphs planting a closing thought in the viewer's mind, didactic titles rounding out the story with historical information (Martin Barker terms these "reality guarantees"),[3] visual roll calls of the cast offering a last look at the characters and the actors playing them, and, in a mediating element specific to this subgenre, lists of the names of the actual soldiers killed in battles depicted in the film (or sometimes other war dead).[4]

In a final look at the PCF subgenre, below I survey the closing narrative moments and the end titles of eleven representative films. Each successive example casts light back on the previous example or examples, putting the subgenre into a kind of conversation with itself. Any film's final expressive moves are tremendously important—as chapter 4's discussion of "God Bless America" in *The Deer Hunter* showed. The cultural stakes for the PCF as an action film subgenre and the cinema as a space for commercial rituals of national identity are brought into sharp relief by the musical endings of the films described below.

Saving Private Ryan's (1998) almost three-hour story ends with a drawn-out narrative cadence worthy of an opera of similar length on which the screen fades to black. After a ten-second silence, an autonomous end-titles cue begins *before* the first title card appears, activating the listener in the blackness. John Williams's end-titles cue—a six-minute, self-contained piece titled "Hymn to the Fallen"—shares no melodic content with the narrative score for the film. Williams also changes the timbral palette, adding a wordless chorus not heard during the narrative score. In short, Williams gives the listening viewer new music that comments on the film just seen (and heard) without quoting from it. The end-titles music encourages the audience to remain in their seats, listen, and reflect. Indeed, the film's press pack prescriptively describes the end titles as a space for reflection, relating how Spielberg was "so moved when he heard the [end titles] music for the first time that he could imagine the audience just sitting . . . listening in a darkened theatre."[5] Posts on America Online suggest that some audiences behaved as Spielberg anticipated: "When the movie ended, no one got out of his or her seat. We all watched as the American flag just waved in the air. Five minutes later [at the end of Williams's "Hymn"], everybody was still in their seat, quiet, except for those with no conscience, who left early."[6]

In its original context, Williams's "Hymn to the Fallen" initiates the proper internal response to *Saving Private Ryan*—articulated elsewhere by Spielberg as "dealing with it privately"—by providing music (in essence, expressively weighty time) in the immediate aftermath of the narrative's close, when the audience ("except for those with no conscience") could sit quietly together and think on what they had seen. "Hymn to the Fallen" is an exemplar of the purpose-composed end-titles cue, a kind of movie music that demands extra time and money to make and gives the composer a chance to comment on the film just seen and the score just heard in a "purely musical" context. The complaint, sometimes made by music scholars, that film music is constrained by the logic of the image track does not apply here—and yet this music is part of the larger film whole. And this "part" easily functions as a "whole," referencing either *Saving Private Ryan* in a concert of Williams's film music, or the larger patriotic values perceived to be expressed by this music made for a serious war film that trumpeted its claim to offer truth about the war and an appropriate multigenerational cinematic experience.

Effectively speaking, films with a narrative cadence and autonomous end titles only nudge viewers toward becoming listeners. Starting the music before the first title card appears is about all this approach allows in the way of shaping audience choices. Williams's stature as a film composer no doubt encouraged some in *Saving Private Ryan's* audiences to stay and listen. Stanley Kubrick similarly nudges the viewer at the close of *Full Metal Jacket* (1987)—although "teases" is perhaps a better description of that film's final musical move. The diegesis ends on Marines marching and singing "The Mickey Mouse Club March." Their voices fade out with no narrative cadence as the image goes to black, as if such music and such soldiers—musically speaking, monstrous children—will go forever marching on.[7] In the blackness, the opening guitar lick of the Rolling Stones' 1966 hit "Paint It Black" breaks the silence. In sync with the start of the drums, the end titles begin with a flash to the words "Directed and Produced by Stanley Kubrick." The titles for *Full Metal Jacket* are cut to the music. As terrific as this little moment is, the game Kubrick plays momentarily pulls the viewer out of the film, with an invitation to identify the track or take immediate pleasure in knowledge of it. The Stones' and Kubrick's audiences would seem to overlap—they do among current male American undergraduates—and the possibility of the viewer affiliating with the track (to use Anahid Kassabian's term for popular music familiar to a movie audience) is enhanced by Kubrick's use of a very well-known band and track.[8] Given the segmented, stylized nature of *Full Metal Jacket's* narrative, punctuated as it is by pop records, "Paint It Black" at the close feels right and a full playing of the record over the end titles affords the viewer who stays a final three minutes of listening pleasure—if they like the Stones. Whether Kubrick's closing musical gambit casts much insight back on the film just seen remains in question. In any event, rhythmic music at the close invites physical rather than internal, reflective response.

Carrying music over the gap between story and credits—building a musical bridge—offers a more robust strategy to turn viewers into listeners. *Platoon* (1986) uses the elegiac strains of Samuel Barber's *Adagio for Strings* to bridge the gap. In its eighth use in the score, the *Adagio* begins to play as a sobbing Chris is flown by medevac helicopter from the battlefield, presumably on his way home (figure 2). The slow-moving music plays on over aerial shots of a mass grave and the jungles of Vietnam as Chris offers specific advice to certain viewers in the last of his many voice-overs: "Those of us who did make it have an obligation to build again, to teach to others what we know, and to try with what's left of our lives to find a goodness and meaning to this life." (The voice-over's prior address to Chris's grandmother

FIGURE 2. A soldier's tear: the invitation to cry at the close of *Platoon*

is tacitly abandoned at the close.) The dense, contrapuntally textured music continues as the screen fades to white for the film's dedication "to the men who fought and died in the Vietnam War." Barber's achingly beautiful, long lines continue their unfolding during a visual roll call of the cast, which gives the viewer a final glimpse of each face, a last moment to think about each character's fate. Indeed, the *Adagio* plays without ceasing to the end of the credits, fading to silence in the middle of the four fortissimo chords that were put to varied expressive use during the narrative, a structurally tense moment in the music that gives no final comfort. In short, *Platoon* never cadences. Stone leaves us in the dark, the *Adagio* hanging unresolved in the air. While the final words of the voice-over suggest the necessity of moving on, the musical postlude expresses nothing but loss.

Platoon demonstrates how an effective musical bridge, in this case with no cadence marking the close of the narrative and, more radically, no cadence at the end of the end titles, can transform viewers into listeners. Interviews with Vietnam veterans who had just seen *Platoon* were common across local and national media. Often these veterans were in tears at the close. Stone built the possibility of tears at the end of *Platoon* into the film by spanning the gap between narrative and end credits with an elegiac musical bridge that denies easy exit from the theater and encourages mournful and sustained reflection on the film, beginning with a last look at the young faces of the cast in the film's final roll call. Barber's *Adagio* here marks the memory of the Vietnam War as an occasion for personal and national mourning. *Platoon* provides the time to remember and music to remember by, creating a ritual of remembrance that many of Stone's fellow Vietnam veterans, as well as general audiences, instinctively participated in. (I did when I saw *Platoon*. I was eighteen years old and had recently completed my selective service registration [draft] card. *Platoon* was, for me and many of my generation, an initial means of access to the "truth" of the Vietnam War. That truth came accompanied by Barber's *Adagio*.) *Platoon*'s original audiences stayed and listened. Many cried. The precise emotional content of those tears, like the expressive content of Barber's piece, remains inaccessible beyond noting the experience of the film as a kind of mourning at once national and personal. Here, once again, *Platoon* sets a benchmark for the cultural work music in the PCF can accomplish.

Barber's *Adagio* at the end of *Platoon* does not take the viewer by surprise. Indeed, for some in the audience repetition of the piece might arouse resistance, with perceived overuse turning elegiac music in its first application into a cliché. But however the *Adagio* is received, the closing music for *Platoon* is of a piece with the film: this music, first heard as Chris and the other "new guys" arrive in "the Nam," reinforces the core aspects of *Platoon* at film's end.

In other PCFs, music at the close works to turn the experience of the whole in a particular direction by spinning out one set of themes to the exclusion of others. Ennio Morricone's score for *Casualties of War* (1989) offers an instance where a composer effectively has the last "word."[9] However, hearing Morricone's final "word" requires the moviegoer to stay and listen and incorporate the score as a full expressive element into any summary reading of the film. Many commentators on *Casualties of War* have not made this musically integrative interpretive effort.

Casualties of War's narrative is framed as a flashback. At the start, Private Eriksson, played by Michael J. Fox, is pictured riding the San Francisco metro. After a young Asian woman sits down in his line of sight, Eriksson falls asleep and returns to the jungles of Vietnam, where he was a soldier. He relives the rape and murder by his squad mates of the young Vietnamese woman Oanh and his efforts to bring the crime to light. After his squad mates are sentenced to prison, Eriksson is jolted awake and the film returns to the opening scene on the metro. Eriksson watches as the young Asian woman exits the train, leaving behind her scarf. (Oanh clung to a scarf throughout her ordeal.) Eriksson rises, collects the scarf, and pursues the girl. Given the imminent end of the film's narrative, the short exchange that follows passes as a highly symbolic conversation. Eriksson hails the girl with a Vietnamese greeting. She turns and asks, "Do I remind you of someone?" He replies, "Yes." She looks at him with sympathy and intuits, "You had a bad dream, didn't you?" Eriksson says, "Yes." The girl finishes the film with the line, "It's over now I think," and offers a quick Vietnamese farewell. She turns and walks away; Eriksson remains. Fox's face, framed by director Brian De Palma in close-up, remains unreadable. The image cuts to the film's final shot: Fox from the waist up, hands on hips, a still-unresolved look on his face, gazing up a bit, apparently unsure what to do, finally exiting the frame, after which the camera cranes up, settling on the distant skyline of San Francisco (a locale that plays no role in the film). The end titles begin here, with the film's title laid over the image of the city. As the image fades to black, the end titles continue.

The girl's parting words—"It's over now I think"—can be read as an incredibly pat, tin-eared conclusion to a graphic, painful film about a still-fresh national trauma; many critics took this view at the film's release. Two examples: "And the final scene, a brief 'redemptive' coda set in San Francisco, is awful—it ends the movie on a phony rhetorical note." And: "'It's over now,' she says. Two million dead, for nothing—that was some dream, all right. Cue the celestial choir; fade out on the TV actor."[10] Even more sympathetic readings of the film—David Greven: "She is giving Eriksson the divine gifts of empathy and recognition, precisely those gifts so resolutely denied to Oanh."—have missed the music.[11] Indeed, attention to the musical content of *Casualties of War*'s narrative close and end titles proves essential to assessing the film as a whole, as at least one critic noted on the film's release: "The heroism of the real-life Eriksson lay not in revenge, but in his devotion to the victim. . . . The leitmotif Morricone uses for Oanh, the mournful pan flute that underscores her plight, comes back like an anguished cry, pushed up past breath, past hope, the last call of the suffering in a world in which death and horror are not only the province of night."[12]

Consider how Morricone's score does important work at the end of *Casualties of War*. The elegiac, consoling theme associated with American soldiers who did not participate in the rape and murder of Oanh starts playing as Eriksson rises from his seat to return the scarf. Previously played for the deaths of Brown and the unnamed "cherry," here it is given to Eriksson. It scores his exchange with the girl in an amplified orchestration with wordless choral voices not heard to this point in the score. After her line "It's over now I think" and the cut to Eriksson's final close-up, the violins make a bold, undeniably melodramatic, entrance, arcing up to the top of their register and shifting the thematic content of the cue toward the modally inflected melodies associated with Oanh. This is the only moment when the score's two primary theme groups are juxtaposed. As discussed earlier, Oanh's themes first sound when she is taken from her village, follow the entire course of her ordeal, and rise to operatic heights at her death. Morricone's specially composed cue for *Casualties of War*'s end titles lasts a further seven minutes: all of it dwells on Oanh's themes.

In his choice of themes and use of a grand choral and symphonic idiom, Morricone directs the viewer-now-listener toward particular aspects of the film. We are enjoined by the composer to think about Oanh, the film's female and Vietnamese victim—virtually the only such character in the entire PCF cycle of the 1980s. Fox's performance suggests that Eriksson is simply unable to process the notion of closure: his exit carries no sense of an ending. Matching this performance, nothing about the final cue suggests that the story just seen or the war just depicted is resolved. Morricone's music, together with the gradual introduction of the end titles, appeals to the viewer to stay in their seat, to not exit the theater, to not assume *Casualties of War* (or whatever is meant in American culture by the word *Vietnam*) is over. One can imagine a moviegoer echoing the final line, saying to a companion, "It's over now I think," and getting the reply, "No, let's stay and listen to the music."[13]

If Morricone's score makes an appeal to the viewer who chooses to listen, director Kathryn Bigelow's interpolated popular music bridge for *The Hurt Locker* (2008) challenges the audience to assimilate her GWOT PCF into the action film genre. *The Hurt Locker* ends with Sergeant James returning for yet another deployment to Iraq. Massive helos taking him back to the battle zone are scored with the huge drumbeats opening "Khyber Pass," a track from the heavy-metal band Ministry's 2006 album *Rio Grande Blood* (the second in a three-disc polemic against President George W. Bush). An exotic mix of pseudo-Arab voices weaves about the noisy musical texture as the helo's rear hatch descends to reveal James among the soldiers inside—a shot not dissimilar from the first appearance of Chris in *Platoon*, only James exits the helo with intention. A shot of James's purposefully striding feet in combat boots, stepping almost in time to the music, cuts to a similar angle on his feet in a bomb disposal suit. The film's final image shows James walking downrange to the track's driving beat, as Stacey Peebles describes the moment, moving "toward yet another puzzle, his face alive and happy, loud heavy metal celebrating him on the soundtrack."[14] Just before the film cuts to black for the start of the end titles, the drums herald a major change in musical texture by way of five fast rim shots. *The Hurt Locker* concludes on images and titles cut to an interpolated metal track.

Reshaping the recording to fit the film's formal needs, Bigelow and her sound editors altered the dimensions of "Khyber Pass." The final cue, a musical bridge, begins at the top of "Khyber Pass" but the five rim shots and the change in texture—coming some three minutes into the original track—land after just one minute and twenty seconds have elapsed in the film. Here musical form, the structure of the track, is adjusted to fit film form, which appears, in film context, to be following musical form. Give and take between image and score drives the editing at the close of *The Hurt Locker.* This alteration to the structure of the track allows Bigelow to end the narrative with a musical bang, evoking, perhaps surprisingly, the aesthetic of more commercial action-adventure subgenres.[15] "Mickey Mousing" the start of the title cards to rhythmically assertive music comes off as an element of genre style rather than an effort to get the viewer to reflect. It's a moment when Bigelow's stated desire to "strike a tonal balance between substance and entertainment" leans toward the latter.[16] Rhythmic music calls attention to the appearance of the cards—perhaps directing us to their content—but more powerfully activates a bodily response. We leave the film moving or just energized, probably speaking loudly, perhaps ready to recommend the film to others. Rhythmically vital end-title cues, especially when drawn from popular music (as in *Full Metal Jacket* and *Three Kings* [1999]), do not encourage the reflective, emotional, or even intellectual response implicit in less forceful music that evokes the concert hall, such as Barber's *Adagio* with its slow, contrapuntal unfolding or Morricone's and Williams's choral and orchestral idioms in *Casualties of War* and *Saving Private Ryan.*

Still, metal at the close of *The Hurt Locker* serves a purpose. This generic action music is heard almost nowhere else in the score and was omitted on the score album.[17] In context, the musical bridge leaves the viewer in a quandary. Few watching will want to return for another tour of duty with the EOD (explosive ordnance demolition) teams profiled in the film, yet James walks out of the film doing just that. It's the only thing left that he "loves," he tells his infant son. In a different action-movie context, Renner's strutting exit would promise a sequel. But the thrill of seeing things blown up and people shot down—fundamental components of Hollywood's business model—doesn't yield much fun for the viewer in this movie. Assuming most Americans watching *The Hurt Locker* don't want to see such a sequel—and not many wanted to see *The Hurt Locker*—Bigelow leaves us with a complicated mix of emotions, triggered powerfully by the music that bridges the viewer into the end titles. As Peebles notes, if at film's end "we move to the heavy beat of the soundtrack and cheer the handsome 'wild man,' we do so by overlooking the nuances of the portrait Boal and Bigelow give us. We miss the point . . . that ultimately *The Hurt Locker* may have as much to say about audiences' addiction to representations of risk as it does about James's obsession with the real thing."[18] There are always, of course, PCF audiences predisposed to "miss the point."

But Peebles exits too soon, for the heavy-metal energy of *The Hurt Locker*'s end titles doesn't last, and those who stay and listen are rewarded with a lyrical, elegiac, mournful musical conclusion more in line with earlier PCFs. The transition to this music unfolds slowly: Ministry's noisy record fades out and a strummed electric guitar chord, a potent totem of white American masculinity, cross-fades in. This chord signals the score's main theme on each of its four appearances in the narrative score. At the end titles, as in the score, the chord establishes a tonal center, a very rare point of conventional musical orientation in a score mostly made of electronic hums, growls, ticks, and fuzz. After several strums, a quavering pedal tone enters in the treble on the pitch that begins the theme itself, hinting that the melody will follow. Bigelow spoke of not wanting to have a score that is "repetitive. You know, you have a phrase, and you know it's going to repeat. Then go onto another phrase. There is no suspense there. Because you know it's going to repeat."[19] In the end titles, suspense is generated in the music by withholding the start of the tune itself. Finally, after much waiting, the theme comes in, scratching an itch built up by the high pedal, the ticking texture, and the score's stingy statements of a mournful movie theme that feels like a pop hook for an American tragedy set in a forbidding land. The return of the lonely, descending theme at the end of the end titles draws the viewer back to the existential plight of the three soldiers at the heart of *The Hurt Locker*. We exit the end-titles "airlock" with music that played when these three confronted the effects of combat on their inner selves ringing in our ears.

Platoon, Casualties of War, and *The Hurt Locker* span the gap between narrative and end titles with a musical bridge. Some scores extend the narrative portion of the bridge far back into the film, creating what I call a long, slow, final movement that activates sustained reflection musically within the film story itself and continues on to provide extended reflective extra-fictional music during the credits. Composer James Horner's score for *Courage Under Fire* (1996) proves an important example because its long, slow, final movement offers both formal and thematic closure denied for the entire previous narrative and score.

The narrative of *Courage Under Fire* concludes with a nine-minute stretch of almost continuous elegiac music accompanying several intercut scenes: a tearful confession (Serling tells Boylar's parents he was responsible for their son's death) and a bittersweet recognition of military sacrifice (the

FIGURE 3.   A soldier's tear: setting down the burden of Vietnam by way of a Gulf War narrative in *Courage Under Fire*

young daughter of the helo pilot Karen Walden has her deceased mother's Medal of Honor placed around her neck in a Rose Garden ceremony). Two further acts of resolution are acted out to elegiac music without dialogue: Serling leaves his Silver Star on Walden's grave marker, then returns to the bosom of his own loving family. In the critic Janet Maslin's words, it's "a 21-tear-duct salute."[20] The only musical break in this long, slow conclusion comes when Serling admits to Boylar's parents that his orders were responsible for their son's death. As the only act of resolution that requires words be spoken, director Edward Zwick faded Horner's music out for a minute, clearing the soundtrack of elegiac music for Serling's confession. (The score album track "The Medal of Honor / A Final Resting Place" includes the music Horner wrote for this minute, scoring Zwick ducked out completely in the mix.) As the elegiac music fades back in, Boylar's father counsels the tearful Serling, saying of his combat mistake, "It's a burden you're going to have to put down sometime" (figure 3). His words function on a larger level as a reference to Vietnam. The film's long musical close helps the viewer, along with Serling, resolve the difficult issues raised by the narrative in a manner that suggests it is time to move on.

Horner's score accomplishes this resolution in a thoroughly musical manner. The film's final elegiac movement begins when Serling's and Walden's stories both reach resolution: Serling comes to a deeper understanding of his own battlefield behavior during the Boylar incident by listening to an audiotape of the battle and, at the same meeting, submits his final report recommending that Walden receive the Medal of Honor. Only at this moment does Horner introduce the score's main themes—a pair of elegiac melodies appearing on the score album in separate tracks titled "Hymn" and "Courage Under Fire." These tracks lay out two broadly stated melodies at similarly moderate tempos, attended by subtly contrapuntal accompanying voices. Both melodies are diatonic, "Hymn" dominated by stepwise motion and "Courage Under Fire" hinting at bugle-call arpeggios with many sigh figures. Each is rounded in form, presenting similar but distinct, dignified elegiac themes.

Horner hints at these hymn-like melodies earlier in the score, especially when Walden is shown away from the battlefield, as a mother with her daughter and when graduating as an officer, but neither is played in its final form until the long, slow, final movement begins. Horner's score struggles for melodic shape just as Serling struggles to reshape himself into the soldier he thought he was while also uncovering the embarrassing series of failures that led to Walden's death. (The fact that she's a woman proves immaterial.) Horner's score withholds the satisfactions of the hymn-elegy style until the truth has been found, a strategy that makes the film's long final cue all the more satisfying. The film story's closing symbolic actions feel more complete for the now-complete melodies accompanying them. And Horner reinforces this sense of closure by marking the end of the narrative with a clear musical cadence, beneath which a pedal tone continues as a musical bridge into the credit roll. As the credits begin, Horner restarts the score's newly found, melodically satisfying title theme, inviting the viewer to stay and listen to music that now, finally, has a rounded, satisfying story to tell.

Elegiac music drawing on the Vietnam cycle of the 1980s and adjusted to lingering issues in the 1990s does substantial work rehabilitating military heroism in *Courage Under Fire*, a self-consciously serious film that refuses to revel in the Gulf War "smart bombs" shown in its opening titles and instead keeps the viewer with the men and women on the battlefield and back home after the war, where enduring human questions remain. Over the course of almost a quarter of an hour of elegiac music, extending across the narrative's close and almost to the end of the end titles, the film's long, slow, final movement marks this serious work of recovery achieved. Having achieved closure, Horner turns back toward heroic action with a return at the very end of the end titles to the action music from the opening tank battle. This is, perhaps, a matter of closing out the score with musical ideas that opened the film, but it also works as an expression of a renewed readiness to go back to war. In absolving military failures—metaphorically

Vietnam—*Courage Under Fire* assures the audience that lessons have been learned and now the fight, if needed, can resume.

Adapting the long, slow, final movement to a GWOT context, *Act of Valor* (2012) closes with a nuanced admission that this war will never end. The final "act of valor" brings the narrative to a momentous conclusion. As the net relays the news that the LT and Chief are both down, the image fades from the wounded Chief to the dying LT—their heartbeats cross-fade one into the other only to fall silent. A hard cut to the LT's flag-draped casket instantly returns his body home. An elegiac cue titled "Engel's Legacy (Funeral)" on the score album accompanies formal military honors. (Engel is the LT's seldom-used name.) With its strings-only orchestration, moderate tempo, contrapuntal texture, and constant quarter-note motion, this cue could just as easily be titled "Barber's Legacy." The match with Barber's *Adagio* is remarkably close. The music briefly pauses for the formal presentation of the flag to the LT's widow, then restarts in the same elegiac vein. Once again, an ABA structure adds a formal dimension to use of the elegiac register by silencing the music for important spoken words, in this case the quasi-liturgical language of the military funeral—"On behalf of a grateful nation . . ." The sounding of "Taps" follows, the apparently diegetic bugler folded into the non-diegetic strings.

The long, slow, final movement continues after the funeral as the Chief—in voice-over—completes the text of his letter to the LT's infant son. As described earlier, halfway through his quoting of a poem by Tecumseh, the LT's voice joins in; then the Chief's voice fades out, leaving the father speaking to the son. The LT's wife lays the letter down before a dual shrine to the family's war dead pairing a World War II soldier who died in 1945 with the LT, dead in 2010—neatly eliding an entire generation of family and national history: Vietnam. Elegiac music plays though all of this, and on into the film's dedication to SEALs killed in combat since 9/11 and "to all of the warriors heading downrange in the future . . ." Behind the dedication and listing of the names of the fallen, the Chief is shown walking alone into the ocean to surf. Near the film's start, the LT taught the Chief to surf. Here, at the close, *Act of Valor* represents the loneliness of those who fought and survived, emphasizing grief as the soldier's lot. Having shown the LT's sacrificial death, attended to the sound of his fading heartbeat, borne his body into the grave with full military pomp, passed a warrior code on to his infant son by way of a voice-over spoken from the dead, and listed the names of actual war dead—all of this to the sound of elegiac music—the long, final movement of *Act of Valor* seems to come to an end. But in fact, it isn't over yet.

Title cards cut to muffled percussive shocks—distant explosions, perhaps the sound of wars yet to come—extol the "damn few" willing to continue "heading downrange in the future . . ." The ellipsis suggests that this battle's end cannot be imagined—at least by the professionals of the all-volunteer military (who would, in such a scenario, be without a war to fight). Here the score cadences and falls silent on a title card reading "Act of Valor." In the silence, "Act of" fades out. Other words fade in with soft, synthesized strings: the screen now offers the epigraph "in Valor there is Hope." Multiple, small images of soldiers, first responders, veterans, and their families begin to fill the screen in collage fashion as the credits begin flashing. The soft music gives way to a guitar intro, to which is added the voice of country singer Keith Urban, who sings the song "For You," composed for the film and the first track on *Act of Valor: The Album*.[21] The song plays in its entirety, the patchwork of images continuing for more than half of its length before the credit roll alone takes over and the viewer is—in effect and finally—released from the film. Urban's clear-eyed lyrics speak of death endured and chosen freely "for you." The focus remains on family. Jingoistic nationalist sentiments, a common feature of patriotic country music after 9/11, are absent.

Act of Valor's very long, slow, final movement, with its extended array of closing tropes, leaves the viewer in an unsurprising place given the extraordinary cooperation accorded the film by the SEAL command and community. Still, it is worth noting the lengths to which the film emphasizes combat deaths as the operative reality for these twenty-first-century warriors, as well as the extent to which music originally employed to mark the moral defeat of Vietnam is put to use consecrating battlefield losses in the GWOT. The sober, drawn-out conclusion refuses any notion that the United States is invincible. Sacrifice "for you" remains the film's final, even fatalistic, certainly resigned, message.

By contrast, *The Kingdom's* (2007) long, slow, final movement—like *The Hurt Locker's* musical bridge but with very different musical content—challenges the action-movie viewer and any moviegoer who embraces military force as a means to win the GWOT. It's a subtle close that the non-listening audience—a big one, to be sure—will likely miss.[22]

Having rescued their team member Adam Leavitt from beheading by terrorists who kidnapped him in a car bomb attack gone wrong, FBI investigators Ronald Fleury and Janet Mayes, together with their Saudi minder Colonel Faris Al Ghazi, enter an apartment in the building where the fight has led them. Janet is lured in by the sound of a young girl crying. In long-standing American soldier tradition, she attempts to calm the girl by offering her a lollipop. The girl responds in turn, with the gift of a marble that

Janet, the team's coroner, recognizes is identical in color and pattern to those she removed from the bodies of bomb victims in the attack that brought the FBI team to Saudi Arabia. Composer Danny Elfman provides a stinger for the marble, which turns the score away from its calm, almost elegiac vibe toward a more threatening veil mood. Discovering they have wandered into the home of the bomb maker Abu Hamza, whom they have been seeking all along, a series of close-range shootings unfold. Hamza is shot—he dies while embracing his young grandson—and so is Faris. Faris dies in Fleury's arms, quickly and graphically, covered in blood from a gaping neck wound, as Fleury whispers, "We got him."

At this point, with the narrative of unambiguous success pulled out from under the film, *The Kingdom* slips into its long, slow, final movement, a six-minute stretch scored by a single elegiac cue. Fleury visits Faris's home for a funeral gathering and has a short conversation with Faris's young son—paralleling Fleury's talk with the young son of his American superior killed in the bomb at the start of the film. Fleury tells both boys their father was his friend. After the funeral, the four FBI agents board a plane for the United States. The entire final sequence unfolds with minimal diegetic sound and sparing dialogue. Music dominates, an elegiac cue marked by a synth-strings background and two electric guitar lines twining about each other.[23] This music continues behind the film's surprising conclusion. Crosscutting between the United States and Saudi Arabia, the same words are spoken by Fleury and the grandson of the bomb maker Hamza. Fleury is asked what he whispered in Janet's ear to comfort and calm her just after the bomb attack early in the film. Hamza's grandson is asked what his dying grandfather whispered in his ear. Fleury replies, "I told her we were gonna kill 'em all." The grandson repeats his grandfather's final words: "Don't fear them, my child. We are going to kill them all." With these words the narrative of *The Kingdom* concludes, fading to black on an extreme close-up of the Arab boy's eyes—a framing that matches similar close-ups of each of the four FBI team members as they leave Saudi Arabia. But while the narrative concludes, the music continues: the melody restarts with the credit roll and plays for a further seven minutes of sustained, elegiac music, including over a memorial title card listing the names of nineteen Americans killed in the 1996 bombing of the Khobar Towers in Saudi Arabia.

Elfman's score is squarely in the beat-driven mode, yet at its conclusion *The Kingdom* shifts definitively toward the elegiac with thirteen minutes of mournful music that refuses to recuperate the violence of the film or send the viewer out on an energetic note. *The Kingdom* ostensibly tells a tale of success in the "war on terror." The terrorist mastermind and bomb maker responsible for the attack that opens the film is killed and the intrepid FBI investigators manage to get the job done despite hurdles erected against them by the Saudis and the US State Department. So, what exactly does the elegy at the close of *The Kingdom* mourn? The immediate narrative focus is the death of Faris, who grows into a real friend of the FBI foursome and emerges as a tremendously sympathetic character. He dies in Fleury's arms—the only foreign ally to get this privilege typically reserved for Americans in the PCF. Unlike Freddy in *Green Zone* (2010), Faris aligns his own interests squarely with the Americans. He is a model example of democratic (as in the rule of law) and pro-US values taking root in the Middle East. That the FBI team mourns his loss so thoroughly counts as a genuine extension of elegiac dignity to a non-American character. Unlike the sadly ironic elegy in *Green Zone* for Freddy's words, "It is not for you to decide what happens here," in *The Kingdom* the foreigner-ally is fully embraced, his loss mourned as if he were American—perhaps the ultimate compliment a Hollywood film can extend to foreign others. But the film's paired final lines—"We will kill them all"—broaden the scope of *The Kingdom*'s verdict on both the story of an FBI investigation and the American approach to fighting terrorism. *The Kingdom* leaves its audience—if they stay and listen—with an emptiness that refuses to be filled with patriotic satisfaction and promises no victory, an opposite final feeling from that afforded by *Act of Valor*.

The power of music at the close, as well as the ease with which audiences can ignore end-titles music simply by exiting the theater or turning off the television, alike spring from the closing of the image track. Most movie watchers assume that when the picture fades, the movie is over. (Film music lovers know better.) To keep the audience engaged, Clint Eastwood includes an eighty-seven-image slide show of authentic photos from the battle of Iwo Jima during the end titles for *Flags of Our Fathers* (2006). History buffs in particular are appealed to from start to finish of the film's very long end titles—more than eight minutes, all scored. The first part pulls gradually away from the main characters swimming on the beaches of Iwo Jima, rising to the heights of Mount Suribachi and a CGI vista of the US armada in the distance. The drawn-out coda to the score's main theme suggests an extended narrative cadence—but the film is far from over. The theme, now in guitar, restarts and the credits continue. Photos to screen left show actual images of the battle of Iwo Jima and the flag raisers' war bond drive while the credits are listed to screen right. For the visual roll call of the cast, photos of the real men substitute for the actors portraying them. So, for example, the last face associated with Iggy is not the actor Jamie Bell at the moment when his name is flashed on-screen but the real Ralph Ignatowski—

the Iggy we never see disemboweled—in a smiling service portrait. During a sequence of graphic battle images, the scoring shifts to surges of metallic sounds, forbidding and disturbing, placed deep in the mix during several battle sequences in the film but heard here without accompanying sound effects—heard, in short, as music. The supplemental quality of the sounds as used in combat scenes is reversed: the metallic sounds alone score the still photos of real combat. With a return to less brutal images for the end of the credit roll, the simple main theme restarts. In musical terms, the end titles are a large ABA structure. The profusion of actual photos—ending a film *about* a famous photo—refocuses the film's splintered narrative. However fractured the storytelling of *Flags of Our Fathers*, actual men lived through these events. The long end-titles slide show insists we attend to documentary evidence for the battle after having endured a cinematic re-creation that at times feels rather abstract. The sentimental music plays on and on, but in a spare manner, without the timbral excess of strings characteristic of the music of *Platoon* or *Saving Private Ryan*. Eastwood keeps it simple: resonant piano and guitar, domestic instruments familiar from daily life. Viewing the images while listening to a plaintive tune and disturbing electronic sounds qualifies as a ritual of remembrance in itself, a postlude with pictures to the act of watching *Flags of Our Fathers*. (The sequence chimes with a short film shown to visitors at the D-Day museum in Caen, France. Documentary footage of the Allies and Germany preparing to clash on the coast of France are shown simultaneously on a split screen, with only music on the soundtrack. The viewer is left to sort out the meaning of the images.) As Holger Pötzsch has noted, "The re-enacted images of [*Flags of Our Fathers*] and the original war footage challenge, question and comment upon each other. In the end it remains up to the audience to combine them into one of various possible meaningful wholes."[24] The two contrasting sorts of music—heard in an ABA structure—direct such "wholes"-seeking remembrance along general lines. The film's waltz-time main theme, in particular, works as a light sentimental glaze on the photos, somewhat like the blue-tinted main titles for *Band of Brothers* (2001).

At the very end of the end titles, diegetic sound returns as Eastwood reopens the diegesis and takes the viewer back to the heights of Suribachi, where a tangle of dog tags left on the flag-raising monument tinkles in the wind. The delicate sound drawn from these metal tabs, generated by the bureaucratic need to identify dead and wounded bodies, adds something new to the sonic lexicon of the combat genre. The return of diegetic sound at the close is itself expressively innovative. No previous PCF makes such a move. Eastwood assumes the audience has remained to the very end of the lengthy slide-show end credits. The subgenre's decades-long history of using the end titles prepared the way for this sequence.[25]

The film's conclusion extends into the end credits the crucial PCF cultural work of displaying the bodies of wounded or dead Americans to the nation gathered together in the secular yet strangely sacred space of the movie theater. (In the twenty-first-century context of federal bans on media photos of flag-draped coffins of American war dead being repatriated, the PCF has served as the only place to see such sights—albeit in a fictional representation.) Brian De Palma's *Redacted* goes a troubling step further.

Redacted closes with a final nod toward De Palma's Vietnam PCF of eighteen years earlier, *Casualties of War*. McCoy, the soldier who didn't participate in—but didn't stop—the rape and murder at the center of the plot, is back home, reunited with wife and friends at a bar. While making of video of the celebration, a friend says, "Tell us a war story." McCoy, like Eriksson when prodded by a chaplain at the EM club in *Casualties of War*, proceeds to recount the events of the film, ending with the line, "And I didn't do anything to stop it." (Eriksson's line: "And I failed, sir, to stop them.") Music begins just as McCoy begins his war story—a lush orchestra-only version of the aria "E lucevan le stele" from Giacomo Puccini's *Tosca*. The effect is startling. This is the first and only scoring cue seemingly activated by a dramatic event. As McCoy falls weeping into his wife's arms, his uncomprehending friends try to recover what one calls "a celebration for a war hero." All clap and the film seems to end on a freeze-frame of the couple: a photo that fades to black. In fact, the Puccini forms a musical bridge, the prelapped sound of the film's final simulated bit of found footage. A title card appears—"Collateral Damage: Actual Photographs from the Iraq War"—followed by the most startling visual roll call in PCF history: images of civilian Iraqi dead, mostly quite small children, many grotesquely bloodied or burned, their eyes blacked out. Cut into the montage are photos of two fictional characters from the film: a pregnant woman who dies as a result of American soldiers firing on her car at a checkpoint, and the young victim of rape and murder (*Redacted*'s Oanh), whose image brings the image track to a close. The music plays over the initial credits and comes to an end. The remainder of the credits flash, then roll in silence.

Redacted's splintered narrative is composed of simulated documentary, media, and Internet sources presented as found footage. De Palma described the inspiration for the montage of photos: "Well, again, this is something I discovered on the Internet. If you put 'Iraqi war casualties' you'll come up with these montages of very sorrowful pictures of, you know, people that have been killed—civilians, with very sad music. And I said, my god, I have got to put this in the movie. This is . . . a form that I discovered on the Internet."[26] (*Generation Kill* concludes on a similar YouTube form borrowed for the purposes of a feature film: a montage of clips capturing the Recon Marines' time in Iraq and accompanied by the Johnny Cash track "The Man Comes Around.") While he obtained releases from the photographers, he was informed by lawyers for the film's distributor that he also needed, in De Palma's words, "releases from dead babies and relatives of dead babies." And so, the blacked-out eyes. The moving close to *Redacted* suggests the lingering power of the PCF as a space to critique American power and the actions of the US military.

Lone Survivor (2013) also uses its closing moments to show photographs and home videos of actual war dead—only here it's the warrior-professionals killed in the depicted story, a failed SEAL mission that ended in the highest one-day battlefield loss of life for the program. Their real names are listed (unlike the living SEALs starring in *Act of Valor*, whose identities are concealed), and the names of the actors who played them are not shown (likely to avoid distraction). This is not a combined real person/credit sequence as in *Flags of Our Fathers* but instead an explicitly commemorative montage, accompanied by the plaintive, haunting sound of Peter Gabriel's 2010 cover of David Bowie's "Heroes," similar to the sort of homemade video montages to music now common at all sorts of public events—graduations, weddings, funerals. To exit the theater when *Lone Survivor*'s story ends and not sit through the ritual of mourning at the close would be unseemly and, for some, explicitly unpatriotic—recalling the people "with no conscience" who left *Saving Private Ryan* before Williams's music concluded.

American Sniper similarly closes with authentic images and video—a montage of Chris Kyle's funeral and posed photos of him and his wife. Perhaps surprisingly, the initial end credits are flashed along with this footage. The music heard here, among the very few cues in the film, is an arrangement of "Taps" interpolated from a 1965 spaghetti Western score by Ennio Morricone. It feels, in context, like a signature Eastwood choice, oddly pulling the PCF, at a most characteristic moment, into a different movie music tradition. As in *Redacted*, once this cue is over the remainder of the credit roll unspools in silence. The audience, held in place by the funeral footage, is forced to leave the theater in uncomfortable silence—as they did on all three occasions when I saw the film.

But the impact of the funeral montage—by this point an almost expected generic move—is undercut by the end of the film's narrative. *American Sniper* leaves out any representation of Chris Kyle's death at the hands of the Iraq War veteran who shot him point-blank at a shooting range, where

Kyle had taken him as part of postwar therapy. Instead of confronting the audience with Kyle's violent death, Eastwood's film closes with Kyle's final minutes at home, where he seems completely rehabilitated, once again the cowboy he was before going to war. The narrative's long, slow, final movement is an unaccompanied idyll of normal family life presented under the dark sign of a title card—"Feb. 2, 2013"—that nods toward the knowledgeable viewer and plays with said viewer's potential dread at having to see Kyle shot. Instead, we only see Kyle depart with the young veteran who will kill him. Taya watches from the door as the screen goes slowly black: the film's only fade-out. As the timpani strokes opening Morricone's cue begin, a title card fades in: "Chris Kyle was killed that day by a veteran he was trying to help." It's a chicken-hawk conclusion. Eastwood's film tries to honor Kyle by removing him from any larger context, just as Kyle does vis-à-vis his own relation to any larger issues of the war in the film. *American Sniper* gives its audience the catharsis of a funeral without showing the trauma of the soldier's broken body. The film's only thoroughly musical moment—its end titles—proves an attempt to slip past the hard issues raised by this unreflective hero's journey, issues raised repeatedly in the narrative and dodged repeatedly by Kyle. The words of the military veteran and scholar Andrew Bacevich apply here: "American warriors may not win wars, but they do perform the invaluable service of providing their countrymen with an excuse to avoid introspection. They make second thoughts unnecessary. In this way, the bravery of the warrior underwrites collective civic cowardice, while fostering a slack, insipid patriotism."[27] *American Sniper*—especially in its close—offers a similar service. In the context of PCF history, Eastwood's film precisely reverses the intentions of the makers of the Vietnam cycle, who desired—in Francis Ford Coppola's words—"to put an audience through an experience—frightening but violent only in proportion with the idea being put across—that will hopefully change them in some small way."[28]

Musical reflection at the close of PCFs creates an opening for audiences to experience the nation in a public setting fueled by the private experience of watching a film (assuming, of course, that these films are experienced in the company of others). As most of my examples have shown, closing music encouraging reflection is a hallmark of PCFs made after the musically seminal *Platoon*, which made strings-only or string-heavy orchestration key to war film scores. This post-*Platoon* reflective music strongly marks the subgenre off from pre-1975 combat films, which virtually all conclude with actual or military-style marches demanding the viewer fall in line and exit the theater in step with the inevitable advance and victory of American power. Such music did not fit frank film narratives of the Vietnam War. Some viewers, of course, resist such nationalistic military music, categorically rejecting all it celebrates. Such viewers are perhaps unlikely to have much tolerance for war movies in the first place. But the elegiac music at the close of PCFs about Vietnam, heard subsequently in films depicting other US conflicts abroad, avoids dividing the audience in this way, instead giving individuals all along the ideological spectrum the opportunity to experience a musical catharsis, affording those who choose to listen the chance to "count the cost" of war in their own way. This funeral music doesn't march us out the door so much as give us the interpretive space and time to decide for ourselves what these films about personal sacrifice for the nation mean. In the case of the PCF, different viewers and listeners will read and hear these film narratives and musical scores in different ways, depending on their interest in film music, their disposition toward the subgenre—which is, after all, driven by depictions of masculinity some find excessive, dangerous, even pathological—and by their personal position within a crucial and fraught category of modern identity: patriotism. For the PCF is, in the end, all about patriotism, a complex cluster of ideas, experiences (many reaching back to childhood), emotions, stories, beliefs, affiliations, and, of course, movies, which together constitute each individual's sense of belonging to the nation. These public and private connections to the idea of the nation take on special energy when narratives of war, soldiers' sacrifice, and the living legacy of veterans are added to the mix, especially for a nation such as the United States, which has been spared almost entirely the experience of total war visited on much of the planet in the modern era—often, of course, by the United States. The viewer's relative social distance from active or retired military personnel and the geographical context in which these films are viewed also affects the personal calculus of patriotism aroused by the subgenre. The immersive space of the cinema, which allows individuals to collectively experience and privately respond to life-and-death narratives of national identity, is a prime place for this kind of personal patriotism to find expression, both for the filmmakers who create these films out of their own needs and for the audiences who voluntarily consume them. Indeed, production and theatrical distribution of these films is surely predicated on a perceived demand for this kind of patriotic cinematic experience, on "the readiness of an audience to cry, to despair, to be wrung out" (as Thomas Larson describes Barber's *Adagio* and other musics of "death and dying").[29]

The PCF—with notable exceptions—seems purposefully designed to activate a tempered kind of patriotism that seeks, not always with success, to avoid caricature or stereotype. PCFs do not end like John Wayne movies,

FIGURE 4.   A soldier's tear: ambiguous victory and nowhere to go at the end of
Zero Dark Thirty

nor does their music sound like music Wayne would have sanctioned. One
historical explanation for this is the subgenre's origin. Born in an impas-
sioned reaction to the Vietnam War by self-identified creative rebels and
archetypal Hollywood liberals such as Oliver Stone, the PCF has in its DNA
a critical, or at the very least nuanced, relationship to military power. The
absorption of *Platoon*'s elegiac music into film scores for movies more sup-
portive of the military or movies telling stories of wars the United States
won should come as no surprise. It's a typical example of mass entertainment absorbing something edgy into the mushy middle. However, there's
reason to hope that end-title music can still open a space for nuanced patriotism in the cinema, as evidenced by a final, post-9/11 example.

Zero Dark Thirty (2012) tells a tale of American power successfully and
surgically applied, re-creating the unbelievably smooth mission to kill Osama
bin Laden. The film manages to remain suspenseful and satisfying even on
repeat viewings, even though the viewer knows how the story will end and
can identify significant signposts. I saw *Zero Dark Thirty* the night it opened
in suburban St. Louis. When the final credits began to roll, a lone woman in
the audience briefly tried to start a chant of "USA, USA," but no one took her
up on it. The end-titles music playing at that moment uses plucked strings—
contrapuntal, equivocal, and unsure in affect—an emptied-out shadow of
Barber's *Adagio*. This music sounds twice at the end of the film. It comes first
at the close of the narrative when the CIA operative Maya, played by Jessica
Chastain, straps herself into a military plane after having identified bin

Laden's body. Maya is the only person on the plane and she remains unresponsive to the pilot's question as to where she wants to go. The plucked music starts as Maya begins to cry—a last, ambiguous act from a character who, for the entire film, knows exactly what she wants (figure 4). With this moment, and to the added sound of sustained strings, the score comes to a somewhat abrupt narrative cadence and the screen cuts to black. Robert Burgoyne describes the final shot this way: "A close up of Maya follows, almost a minute in length, a portraiture shot that slows the film to a halt, as the audience is invited to take stock, to consider the moment, and to reflect."[30] When the plucked music starts up again in the blackness, activating the viewer's ear just as in *Saving Private Ryan*, the audience is invited by the score to continue considering the moment, to reflect further on the film, to leave, perhaps, in silence, or to remain, perhaps, in silent thought.

Zero Dark Thirty closes with no resolution in sight and the initial end-titles music leaves the viewer-now-listener hanging. Director Kathryn Bigelow and composer Alexandre Desplat keep the PCF movie music tradition alive. We are invited to reflect on the successful killing of bin Laden, but given no musical cue to feel like anything much was accomplished by this act. Maya never answers the pilot's question and the film's final line: "Where do you want to go?" Audience members might repeat this question to each other in the more immediate social context of a trip to the movies, or return to the final line of *Casualties of War*, turning it into a question that applies to *Zero Dark Thirty* and the war it depicts, asking unsurely (but, alas, wrongly), "It's over now, I think?"

Acknowledgments

Material from this book was presented at the University of Texas at Austin GAMMA-UT conference in 2013, the American Musicological Society in 2013, Music and the Moving Image in 2014, and the Society for Cinema and Media Studies in 2012 and 2014. Anna Froula and Stacy Takacs provided a chance to expand my thoughts on *Band of Brothers* in a chapter for the book American Militarism on the Small Screen (Routledge, 2016).

A summer fellowship from the Dean of Arts and Sciences at Washington University in St. Louis supported archival research in Hollywood. Susan McClary and Thomas Schatz endorsed an early proposal of the topic. Robert Burgoyne generously shared his work on *Zero Dark Thirty* prior to publication. Cindy Lucia helped me make a valuable contact. Gaylyn Studlar reminded me to stay attuned to traces of sentimentality.

Students in my courses on the combat film and film music and sound helped me watch and listen in new ways. I am especially grateful to Caleb Boyd, Adam Caplan, Sami Lavin, Marc Niemeyer, Erin Sellers, Eve Sembler, Ted Sorota, and Spencer Welsh.

My original editor, Mary Francis, nurtured this book over many years. It was, once again, a pleasure and an honor to make a book with her on my side. My subsequent editor, Raina Polivka, and the anonymous readers for UC Press made substantive critiques that improved the book tremendously. Many thanks to Lindsey Westbrook for expert copyediting and to Zuha Khan for seeing the manuscript through from submission to production. Jennifer Psujek helped with research. Dan Viggers made the musical example. Caleb Boyd and Ashley Pribyl assisted with fact checking.

My interest in war movies began with a film that, sadly, only made it into the endnotes for this book: Roland Joffé's *The Mission* (1986). I saw *The Mission* when it came out in theaters with my then-girlfriend, now wife, Kelly. Our shared conversations about movies and war and peace began back then and continue to the present. I can't imagine writing this book without her in my life.

My younger son, Jamie, sat next to me on the couch reading the Hardy Boys while I wrote and edited this book. Time together reading with him has been a precious part of this project.

My older son, David, watched almost all the movies discussed in this book together with me. Some, like *Band of Brothers* and *Saving Private Ryan*, we shared several times. It is no exaggeration to say that this book was written in large part so that David and I could talk movies and writing together. I will miss that aspect of this project most of all. My fondest memory of working on *Hymns for the Fallen* comes from the afternoon David and I saw *Lone Survivor* in the theater. When the end titles were over—we always watch to the end of the end—and I began to think about how I would have to incorporate yet another film into the project, David looked at me and said, with enthusiasm, "Dad, this movie will be great for your book." This book is for him.

Abbreviations

Notes

INTRODUCTION

1. Scruggs and Swerdlow 1985, 7.
2. Hinson 1987.
3. *HH* script, author's collection.
4. Stanley Kauffmann, "Don't Mean Nothin'," *New Republic*, September 1987; Duane Byrge, *Hollywood Reporter*, 5 August 1987.
5. Savage 2009 [2011], 4.
6. Schubart and Gjelsvik 2013, 7.

CHAPTER ONE

1. Appy 2015, 273. The *Rambo* (1982, 1985, 1988, 2008) and *Missing in Action* (1984, 1985, 1988) franchises alike explored the persistent fantasy of refighting the Vietnam War. *Uncommon Valor* (1983) took up similar themes with a bit more seriousness but still offered viewers unambiguous victory at the close. Later wars would inspire similar cycles of unreflective combat films: for example the *Sniper* (1993, 2003, 2004, 2011, 2014, 2016 [direct to video]) and *Behind Enemy Lines* (2001, 2006, 2009, 2014 [a SEAL Team film]) franchises.
2. Cooper 1988, 76.
3. Jay Winter, "Filming War," in D. Kennedy 2013, 158.
4. McManus 2010, 63.
5. Cooper 1988, 76.
6. Savage [2009] 2011, 20.
7. Hellmann 1986, 221.
8. Kinney 2000, 135.
9. Herr 1977 [2009], 45.
10. Kieran 2014, 3.
11. Aveyard and Moran 2013, 11.
12. Final visual roll calls are found in *The Deer Hunter, Platoon, Hamburger Hill, 84 Charlie MoPic, Born on the Fourth of July, Windtalkers,* and *Jarhead.*

13. Harper Barnes, "An Action Flick for Thinking People," *St. Louis Post-Dispatch*, 12 July 1996.

14. *Windtalkers* press kit, AMPAS.

15. Lawrence Van Gelder, "At the Movies," *NYT*, 28 August 1987.

16. Jay Sharbutt, "Proposed Vietnam Film: A Real Battle Written by a Veteran," *LAT*, 9 February 1985.

17. Arwa Haider, "War of Words," *Time Out*, 1–8 March 2000.

18. Glenn Whipp, "Intensely Focused," *LAT*, 23 December 2009.

19. Cowie 2001, 113.

20. *AN* press kit, AMPAS.

21. *AN* premiere program, AMPAS.

22. Bill Higgins, "*Ryan* Leaves Them Speechless," *LAT*, 23 July 1998.

23. Kornbluth and Sunshine 1999, 20, 55, 93, 99, 82.

24. John Pilger, "The Gook-Hunter," *NYT*, 26 April 1979.

25. "Ridley Tells Why War Film's Premiere Was Brought Forward," *Journal* (Newcastle), 18 January 2002.

26. Geoffrey Gray, "Activists Protest No. 1 Movie," *Village Voice*, 12 February 2002.

27. Nicholas Kristof, "The Wrong Lessons of the Somalia Debacle," *NYT*, 5 February 2002.

28. Suid 2002.

29. *SPR* "final shooting script," AMPAS.

30. *SPR* press pack, AMPAS.

31. Ambrose 1997, 471.

32. Kristin Hohenadel, "Learning How the Private Ryans Felt and Fought," *NYT*, 17 December 2000.

33. Richard Huff, "Actors and Vets Bond in *Band of Brothers*," *Daily News*, 9 September 2001.

34. *USA Today*, 18 January 2002.

35. Thomas Quinn, "Interview Ridley Scott," *Mirror*, 18 January 2002.

36. *We Were Soldiers* press kit, AMPAS production file.

37. Bowden 2002, xi–xii.

38. *Green Zone* press kit, AMPAS.

39. Simi Horwitz, "Not Easy Bein' *Green*," *Backstage*, 18–24 March 2010.

40. Several earlier PCFs used real amputees in action scenes, easing the technical challenge of showing the effects of combat on the bodies of background characters. In addition, the Marine boot camp drama *The D.I.* (1957) cast active-duty soldiers as primary characters.

41. Purse 2011.

42. Jean Oppenheimer, "Close Combat: HBO's Intense 10-Part Miniseries *Band of Brothers*," *American Cinematographer* (September 2001); *Jarhead* press kit, AMPAS.

43. Peter Rainer, "A Few Good Men," *New York*, 17 June 2002.

44. Purse 2013, 2.

45. Philip Drake, "Distribution and Marketing in Contemporary Hollywood," in McDonald and Wasko 2008, 64.

46. Charles Acland, "Theatrical Exhibition: Accelerated Cinema," in McDonald and Wasko 2008, 86.

47. Buzzell 2005, 73.

48. Wright 2004, 195, 206.

49. Swofford 2003, 9, 7–8.

50. Chapman 2008, 128. Chapman makes this distinction in reference to the post–World War I cycle of silent films that presented "the futility and waste of war."

51. Chapman 2008, 184; see also 243–44.

52. Stubbs 2013, 13.

53. *Go Tell the Spartans* and *84 Charlie MoPic* were both shot in Southern California. With a budget under $1 million, *Go Tell the Spartans's* star, Burt Lancaster, already accepting a much lower fee than normal, ended up putting $150,000 of his own money into the project (Suid 2002, 348–51). *The Boys in Company C* was financed by the Hong Kong producer Raymond Chow, and *The Deer Hunter* by British Lion, later absorbed by EMI (although the film's US distributor, Universal, provided some production funds). Oliver Stone described *Platoon*, produced by the British-based Hemdale Films, as made "with European money" and decidedly "not a major studio film" (Stone 1987, 6, 11).

54. Paul Rosenfeld, "How *Casualties of War* Survived," *LAT*, 13 August 1989.

55. Steve Persall, "War: What's It Good For? Maybe a Few Good Laughs," *St. Petersburg Times*, 1 October 1999.

56. Suid 2002, 612.

57. *Spoils of War* screenplay, author's collection.

58. Peter M. Nichols, "*Three Kings* in Hindsight," *NYT*, 7 April 2000.

59. Hammond 2002, 72.

60. Murch 2001, 4.

61. The *Ryan-Line* cycle may have closed. Writer-director David Ayer's 2014 tank-centered PCF *Fury* perhaps marks a fresh direction. This claustrophobic film is difficult to categorize as a PCF. The crude, hardened tank crew at the center of the plot repeat as their mantra the phrase "best job I ever had," and they go down fighting, saying of their tank, "This is my home." Drenched in the muck of combat and decidedly not oriented toward generational obligation, *Fury* innovatively applies, somewhat harshly, a warrior ethos to the men who fought World War II.

62. Kim Masters, "Against the Tide, Two Movies Go to War," *NYT*, 4 November 2001.

63. Glenn Whipp, "Intensely Focused," *LAT*, 23 December 2009.

64. Anne Thompson, "Big Directors Turn to Foreign Investors," *Variety*, 11 September 2008.

65. John Horn, "An Iraq War Movie with a Bourne Feel," *LAT*, 27 February 2010.

66. Barker 2011, 40.

67. Quoted in Kieran 2014, 174.

68. Mark Boal, "Death and Dishonor," *Playboy*, May 2004, 108–12, 134–41; and McCain 2010.

69. Crouch and Galdorisi 2012, ix.

70. Drone warfare is not especially cinematic. As Air Force General Michael Ryan has said, "[Drones] go out there and die for their country and we don't mourn" (Arkin 2015, 67). Only two films to date have taken up the subject in a serious vein: *Good Kill* (2015) and *Eye in the Sky* (2016).

71. Bergen 2012, 152.

72. Cockburn 2015, 2, 245.

73. Richard Schickel, "Soldiers on the Screen," *Time*, 17 December 2001.

74. Mike Clark, "*Black Hawk Down* Turns Nightmare into Great Cinema," *USA Today*, 28 December 2001.

75. "Press Conference for *Zero Dark Thirty*, 4 December 2012," in Keough 2013, 226.

76. Arkin 2015, 100.

77. The SEALs who arrive near the end of Paul Greengrass's *Captain Phillips* (2013) to expertly conclude a hostage crisis are similarly opaque.

78. The 2016 film *13 Hours*, about military contractors—retired SEALs, mostly—who fought and died during the 2012 attack on the American consulate in Benghazi, is not included among PCFs here. The film does not represent commissioned American soldiers or the US military per se, and so its appeal to patriotic sentiment is complicated. See Hagedorn 2014 for more on military contractors.

79. Clancy 2012, xi.

80. Robert L. Goldich, "American Military Culture from Colony to Empire," in D. Kennedy 2013, 98–99.

81. Andrew Sarris, "A Reality-Based Epic of Bravery and Futility," *New York Observer*, 14 January 2002.

82. Basinger 1986 [2003], 10, 5.

83. Mikkel Bruun Zangenberg, "Humanism versus Patriotism?: Eastwood Trapped in the Bi-Polar Logic of Warfare," in Schubart and Gjelsvik 2013, 219, 222.

84. Vibeke Schou Tjalve, "To Sell a War: Flags, Lies, and Tragedy," in Schubart and Gjelsvik 2013, 252.

85. Bailey 2009, 249.

86. Wong 2005, 610.

87. *Flags of Our Fathers* invokes the imperative "leave no man behind" ironically in the context of World War II. As the massive armada of ships steams its way to Iwo Jima, a marine falls overboard. The men of the story watch and wait for him to be rescued. When it's apparent the convoy will not stop, one mutters, "So much for no man left behind."

88. Slavoj Žižek, "Green Berets with a Human Face," *London Review of Books* blog, 23 March 2010, http://www.lrb.co.uk/blog/2010/03/23/slavoj-zizek/green-berets-with-a-human-face/.

CHAPTER TWO

1. *The Apocalypse Now Sessions: The Rhythm Devils Play River Music*, Passport Records, 1980.

2. Jackson 1979.

3. Cristy Lytal, "His Sounds Are Loud and Clear," *LAT*, 10 January 2010.

4. Koppl 2009a.

5. Rudy Koppl, *"The Hurt Locker*: Touched by an Angel," musicfromthemovies.com, posted 2009, http://www.musicfromthemovies.com/index2 .php?option=com_content&view=article&id=63.

6. Théberge 2008.

7. Dupuy and Bregstein 1946, 76.

8. Castner 2012, 194.

9. Gray 1959, 12.

10. Amy Taubin, "Hard Wired: Kathryn Bigelow's *The Hurt Locker*," *Film Comment* 45, no. 3 (2009): 31–32, 34–35.

11. Scott Tobias, "Interview: Kathryn Bigelow (AVClub.com, 2009)," in Keough 2013, 158.

12. Pelfrey 1987, 157.

13. Quoted in Carrie Rickey, "Return Engagement," *Philadelphia Inquirer*, 13 January 2002.

14. Glenn Man, "Clint Eastwood's Postclassical Multiple Narratives of Iwo Jima," in Schubart and Gjelsvik 2013, 146.

15. Kassabian 2001, 3.

16. An episode 7 visit to a motor pool briefly features a hip-hop track on a portable radio. The Recon Marines at the center of the series do not have such audio equipment.

17. Tom Tunney, "Over the Hill," *HH* clippings, AMPAS.

18. *SPR* press kit, AMPAS.

19. *Saving Private Ryan: Music from the Original Motion Picture Soundtrack*, DreamWorks DRMD-50046, 1998.

20. For analysis of this moment, see Inglis 2005.

21. Psujek 2016.

22. Composer James Horner's score for *Windtalkers* is old-fashioned in both its use of a conventionally recorded orchestra and its lavish length and point-to-point scoring of action scenes and moments of reflection. The score accompanies the full length of most battle sequences—unusual for this style of music—but is hard to hear much less follow given its low prominence in the mix relative to effects. Terence Blanchard's score for *Miracle at St. Anna* plays almost continuously, with the ethnic comedy of the Italian villagers often leading to quasi-comic, stereotyped scoring—such as unfortunate pizzicato strings and accordions for a bumbling old Fascist.

23. *B4J*, first two quotes from draft script dated 1978, third from script dated 1988, AMPAS.

24. Williams would go on to score Stone's *JFK* (1991) and *Nixon* (1995).

25. *B4J*, 1988 script, AMPAS.
26. Dienstfrey 2016.
27. Jackson 1979.
28. "Walter Murch" 1996, 152–53.
29. Daughtry 2015, 3.
30. Daughtry 2015, 11.
31. Daughtry 2015, 23, 29.
32. Daughtry 2015, 11.

CHAPTER THREE

1. Tregaskis 1943 [2000], 4.
2. Tregaskis 1943 [2000], 161.
3. *Generation Kill* is also distinguished by its frank treatment of soldiers' sexual lives, specifically masturbation. The episode titled "Combat Jack" treats the activity as an expected part of life on the battlefield; another includes a masturbating soldier, albeit blurry and unidentified, observed by Wright late one night. *Jarhead*, the only other PCF alluding to or representing masturbation, links the activity to Swofford's suspicion that his girlfriend back home is cheating on him. Typically the PCF, like most combat novels and memoirs, shields the viewer from such surely ubiquitous soldier activities.
4. Lichtenfeld 2007, 1.
5. See http://gawker.com/the-top-100-greatest-action-movie-one-liners-the-super-1486779271.
6. Stone 1987, 60.
7. Barker 2011, chapter 2.
8. West 2005, 150, 279, 290. All of chapter 26 is devoted to the pleasures of battle.
9. Wright 2004, 2.
10. Gregory Weinkauf, "War Is Heck," *New Times LA*, 30 September 1999.
11. *SPR* original screenplay, author's collection.
12. Greengrass 2006, 49, 69.
13. Quoted in Kieran 2014, 166–67.
14. Bradley 2000, 189.
15. Chapman 2008, 64, quoting Bosley Crowther on *Wake Island*.
16. Bradley 2000, 308.
17. SK/16/2/16.
18. Nolan 2002, 3.
19. Nolan 2002, 5, 124.
20. *Black Hawk Down* script "11/7/00 (D.D.)," AMPAS.
21. Nolan 2002, 31, 112, 126.
22. Dave Pomeroy, *Film Info*, undated, *The Boys in Company C* clippings, AMPAS.
23. *Variety*, 1 January 1978.
24. *Box Office*, 9 January 1978.

25. McQuiston 2013, 13.
26. Jack Kroll, "1968: Kubrick's Vietnam Odyssey," *Newsweek*, 29 June 1987.
27. SK/16/1/3/21.
28. Sean Mitchell, "Vietnam Vet Is Star Recruit for *Jacket*," *Los Angeles Herald-Examiner*, 26 June 1987.
29. *HH* script, author's collection.
30. Modine 2005, 199.
31. Bradley 2000, 70.
32. Hasford 1979, 11.
33. SK/16/2/16; Buchanan 2007.
34. Brennan 1987; Caputo 1977; Ford 1967.
35. Bacevich 2013, 9.
36. Jack Kroll, "Remembering Hamburger Hill," *Newsweek*, 14 September 1987.
37. Hinson 1987.
38. Huebner 2008, 233.
39. Marlantes 2010, 428–29.
40. Variations on "It don't mean nothin'" resound in later films. In *Black Hawk Down* Eversmann and Smith exchange the words "This is nothin'" as the latter dies. The refrain "nothing at all" moves across *American Soldiers*.
41. Luttrell 2007, 243.
42. Finkel 2013, 111.

CHAPTER FOUR

1. Basinger 1986 [2003], 71.
2. Tregaskis 1943 [2000], 4, 15.
3. Jorgenson 1991, 36, 183.
4. Daughtry 2015, 227.
5. Pieslak 2009, 160.
6. *HH* script, author's collection; press kit, AMPAS.
7. *Platoon* script, 5 June 1984, author's collection.
8. *Platoon* script, 10 February 1986, AMPAS.
9. See Decker 2017b for further discussion of the song scene.
10. Stone 1987, 32.
11. Janet Maslin, "Fighting the Battle of Money and Greed," *NYT*, 1 October 1999.
12. *Three Kings* script "6/22/98," AMPAS.
13. *Jarhead* script, AMPAS.
14. *AN* script, WGFL.
15. *AN* script, WGFL.
16. *AN* script "1975," AMPAS.
17. *AN* script, "1975," AMPAS.
18. *AN* script, WGFL.

19. *AN* script "1969," AMPAS.

20. *AN* script "1975," AMPAS.

21. *AN* script "1975," AMPAS.

22. *AN* script "1969," AMPAS.

23. *AN* script "1975," AMPAS.

24. The men of Easy Company do sing some: a snatch of "Blood Upon the Risers" (alternate lyrics to "The Battle Hymn of the Republic" using the phrase "Gory, Gory, What a Hell of a Way to Die") in nice harmony, a bit of "Oklahoma!" at mention of the Broadway show, a shred of "I Dream of Jeannie" shushed out in the snows of Bastogne, and a fragment of the real Easy Company's "Curahee Running Song" (a nod toward insider military culture caught, like "Ballad of the Frogman" in *Lone Survivor*, only by those in the know). These songs—except for "Oklahoma!," which facilitates the mocking of a replacement—mostly locate the men of Easy Company in military rather than American popular culture.

25. *Flags of Our Fathers* script, author's collection.

26. Daughtry 2015, 220.

27. Cusick 2008.

28. Buzzell 2005, 207–8.

29. The actual title of the Drowning Pool song is "Bodies." No soldier refers to it in this way in the published literature on the war.

30. West 2005, 176–77, 273, 275, 173.

31. Daughtry 2015, 57.

32. Buzzell 2005, 77–78, 73.

33. Gilman 2016, 2.

34. Wright 2004, 21, 77.

35. Wright 2004, 152–53.

36. K. Kennedy 2010, 47.

37. Engel 2008, 297–98.

38. Pieslak 2009, chapter 2.

39. Pieslak 2009, 150.

40. Daughtry 2015, 141.

41. Lance Morrow, "Viet Nam Comes Home," *Time*, 23 April 1979.

42. Arthur Knight, *Hollywood Reporter*, 1 December 1978.

43. Charles Champlin, "Deer Hunter: A Palship Goes to War," *LAT*, 3 December 1978.

44. Jack Knoll, "Life-or-Death Gambles," *Newsweek*, 11 December 1978.

45. Marc Green, "*Deer Hunter*: the War Hits Home," *Chronicle of Higher Education*, no. 8 (January 1979).

46. No author, "On Film" *Westways*, January 1979.

47. Peter P. Schillaci, "Exorcising the Demons of Vietnam," *Christian Century*, no. 21 (March 1979).

48. Stanley Kauffmann, "The Hunting of the Hunters," *New Republic*, 26 May 1979.

49. Frank Rich, "In Hell Without a Map," *Time*, 18 December 1978.

50. Jay Scott, "Deer Hunter: A Dissonant Symphony for a Lost Cause," *Globe and Mail*, 17 February 1979.

51. *The Deer Hunter* "Location Draft," 28 January 1977, AMPAS.

CHAPTER FIVE

1. Appy 2015, 81.

2. "Walter Murch" 1996, 162.

3. *B4J* script, author's collection.

4. Cockburn 2015 discusses the unreliability and low resolution of live combat feeds.

5. Pelfrey 1987, 5.

6. The novelization of *Act of Valor* does celebrate such equipment (Crouch and Galdorisi 2012, 98). Daughtry 2015, 145–46, expands the list of contemporary hearing technologies not represented in film, including the Combat Arms Earplug, Combat Vehicle Crewman Headset (CVC), and QuietPro. In such prosthetics, "loud impulse noises" trigger "robust noise reduction," shielding and protecting the soldiers' hearing, rendering "their wearers auditory cyborgs of sorts, with ears more robust and sensitive than those of unassisted auditors."

7. Daughtry 2015, 80.

8. Dye is a fixture in the Vietnam and World War II PCF cycles. Oliver Stone hired him as a military advisor and to run a preproduction boot camp for the cast of *Platoon*. Dye would expand this opportunity into the company Warriors Inc., which assists Hollywood film productions on technical aspects of representing the military on-screen. Eight PCFs are listed as company projects on Warriors Inc.'s website. Dye appears in *Platoon* (his second role) as well as *Casualties of War*, *84 Charlie MoPic* (as a radio voice), *Born on the Fourth of July*, *Saving Private Ryan*, and *Band of Brothers*, always playing officers a step or so higher on the chain of command than the main characters.

9. Stone 1987, 111.

10. Cowie 2001, 103.

11. Peebles 2011, 159–60.

12. *HH* script, author's collection.

13. *Miracle at St. Anna* posits a battle space with a live voice-over. A loudspeaker on a German truck broadcasts the voice of an Axis Sally–type woman who hurls racially charged language at African American soldiers approaching the German front line. The black soldiers mostly ignore her words, which are more for the film audience. The image track cuts to the woman in a stagy Berlin radio booth, her physical performance—complete with long cigarette holder—pushing the moment toward campy shtick. Still, it works as a way to articulate racist ideologies on both sides, with the grotesquerie of Nazi propaganda critiquing the realities of American racism.

14. Chion 1999, 51.

15. Chion 1999, 54.

16. Polan 2004, 59.

17. Pfeil 2004, 181.

18. Chion 1999, 23.

19. *SPR* script, author's collection.

20. Lars-Martin Sørensen, "East of Eastwood: Iwo Jima and the Japanese Context," in Schubart and Gjelsvik 2013, 207–8.

21. Chion 1999, 47.

22. The full text of Marc Lee's letter is posted at http://americasmighty warriors.org/marcs-last-letter-home/.

23. Michelle Lee, "*American Sniper* Portrayal Hides the Real Marc Lee," opinion piece on the website Foxnews.com, posted 20 February 2015, http://www .foxnews.com/opinion/2015/02/20/american-sniper-portrayal-hides-real-marc-lee.html; Melissa Clyne, "Mother of Kyle's Fellow SEAL: Portrayal in Film 'Was Not My Son,'" Newsmax.com, posted 3 February 2015, http://www.newsmax .com/Newsfront/Marc-Lee-Debbie-Lee-Clint-Eastwood/2015/02/03/id/622383/.

CHAPTER SIX

1. *The Thin Red Line* shooting script, AMPAS.

2. Nolan 2002, 41.

3. Daughtry 2015, 76, 80, 88, 92.

4. Hemphill 1998, 17, 83, 169, 192, 83, 65, 117.

5. Turner 2005, 9.

6. Hasford 1979, 96.

7. Nolan 2002, 49, 54, 106.

8. Finkel 2013, 5.

9. Junger 2010, 151.

10. Chapman 2008, 84.

11. Finkel 2009, 254–55.

12. Jack Anderson and Dale Van Atta, "The Soul of Gen. Schwarzkopf," *Washington Post*, 24 February 1991.

13. Pauline Kael, "The God-Bless-America Symphony," *New Yorker*, 18 December 1978.

14. John Powers, "Pure War," *LA Weekly*, 18 January 2002.

15. I do not, however, dwell analytically on how PCF soundtracks deploy the surround speakers for purposes of immersion.

16. Music comes in for the final phase, when Basilone exposes himself to enemy fire to clear a pile of Japanese dead, giving the Americans an unobstructed field of fire, and when he runs to get more ammunition.

17. Daughtry 2015, 92.

18. Burgoyne 2008, 53.

19. Malick lays out a similar shot reverse shot layout in *The Thin Red Line* between the Americans' position at the bottom of a hill and the Japanese machine gun in a bunker they must take out at the top. Having taken the bunker, Gaff blows a whistle to signal to those below that the objective has been secured.

20. Pelfrey 1987, 118; Hasford 1979, 126.

21. Stone 1987, 123.

22. Leckie 1957, 214.

23. Robinson 2013, 11.

24. Peebles 2011, 3.

25. Turner 2010, 6.

CHAPTER SEVEN

1. Suid 2002, 356

2. Peter Arnett, "*The Deer Hunter*: Vietnam's Final Atrocity," *LAT*, 8 April 1979.

3. Caputo 1977, 75–76.

4. Zaffiri 1988, 9.

5. Buzzell 2005, 348–49.

6. Pelfrey 1987, 76.

7. SK/16/2/8/1–2.

8. SK/16/4/2/5.

9. Jack Kroll, "Remembering Hamburger Hill," *Newsweek*, 14 September 1987.

10. "Walter Murch" 1996, 153.

11. *The Deer Hunter* "Location Draft," 28 January 1977, AMPAS.

12. Hinson 1987; J. Hoberman, "Hollywood on the Mekong," *Village Voice*, 8 September 1987.

13. Springsteen, keynote speech for South by Southwest music festival, 15 March 2012. In Burger 2013, 391.

14. Hillstrom and Hillstrom 1998, 291, 294.

15. Unattributed clipping, Tom Tunney, "Over the Hill," *HH* clippings file, AMPAS.

16. Herr 1977 [2009], 150.

17. Nolan 2002, 35; Phillip McCarthy, "Combat Rock," *Sydney Morning Herald*, 15 February 2002.

18. Lichtenfeld 2007, 68.

19. Bailey 2009, 60 (both quotes).

20. Bergen 2012, 167.

21. *Green Zone* script, WGFL.

22. Bergen 2012, 183.

23. McChrystal 2013, 165.

24. Bergen 2012, 220.

25. Owen 2012, 207.

26. Owen 2012, 205.

27. Bergen 2012, 228.

28. Just before the explosion—as the other helo is loaded with bin Laden's body and lifts off—Desplat activates a hurry beat unlike any heard in the film so far, a rare moment of added musical energy in the film.

29. The four explosions could also be read as four separate charges going off in turn.
30. *AN* script "1969," AMPAS.
31. Pieslak 2009, 85.
32. Murch 2001, 2–4.
33. Cowie 2001, 102.
34. "Walter Murch" 1996, 158.
35. Cowie 2001, 39.
36. Friedman n.d.
37. *AN* script "1975," AMPAS.
38. *AN* script "1975," AMPAS.
39. *AN* script "1975," AMPAS.
40. *AN* script "1969," AMPAS.

CHAPTER EIGHT

1. Suid 2002, 117.
2. O'Brien 2012, chapter 2.
3. Appy 2015, 124.
4. Huebner 2008, 175, 238, 243.
5. Huebner 2008, 103, 108.
6. *Platoon and Salvador: Original Motion Picture Soundtracks,* composed and conducted by Georges Delerue, Prometheus Records PCD 136.
7. Alternately, the original elegiac tag had to be cut, as Barber's *Adagio* had replaced Delerue's elegy theme by this point in the making of the score (see chapter 10). In any event, the Barber unfolds too slowly to effectively fill a mere minute of screen time, and it never flows out of or into Delerue's veil cues. The original scoring and compiled cues in *Platoon* remain completely segregated.
8. Alexander Walker, "Stanley Kubrick's War Realities," *LAT,* 21 June 1987.
9. Gengaro 2012, 62.
10. Koppl 2009a.
11. Rudy Koppl, "The Hurt Locker: Mainlining the War at Death's Door." musicfromthemovies.com, posted 2009, http://www.musicfromthemovies.com/index2.php?option=com_content&view=article&id=64.
12. Koppl 2009b.
13. Koppl 2009b.
14. Koppl 2009b.
15. Filkins 2008, 116.

CHAPTER NINE

1. McNeill 1995.
2. A less-than-subtle added text for Kamen's tune, titled "Requiem for a Soldier" and sung by soprano Katherine Jenkins at a 2007 concert in London's Royal Albert Hall in tribute to British troops fighting in the Middle East,

reveals an effort to assign specific ideological meaning to Kamen's sentimentally vague waltz. (Jenkins's performance is usually available on YouTube.) The stilted, cliché-ridden text speaks from the perspective of the present and addresses soldiers fallen in war, beginning, "I wish you'd lived to see / All you gave to me." The lyric incorporates the phrase "band of brothers"—a bit of brand placement, albeit drawn originally from Shakespeare—and attaches the idea of fighting for "one shining dream of hope and love / life and liberty" directly to the soldiers themselves, describing a day when all, including soldiers fallen in war, will "live together, when all the world is free." The Albert Hall version reassigns Kamen's tune—a piece of pre-9/11 popular culture tuned to a highly specific exploration of the "Greatest Generation"—to the urgent ideological needs of the US and British coalition fighting, and at the time losing badly, in Iraq and Afghanistan.

   3. Caryn James, "Intricate Tapestry of a Heroic Age," *NYT*, September 7, 2001; Todd McCarthy, Television Reviews, *Variety*, September 4, 2001.

   4. See Decker 2016.

   5. Thomas Schatz, "Old War / New War: *Band of Brothers* and the Revival of the WWII War Film," *Film and History* 32, no. 1 (2002): 77.

   6. Kieran 2014, 154.

   7. Gina Piccalo and Louise Roug, "Scoring, Post-Terrorism," *LAT*, 17 January 2002.

   8. George Monbiot, "Both Saviour and Victim," *Guardian*, 29 January 2002.

   9. Gina Piccalo and Louise Roug, "Scoring, Post-Terrorism."

   10. See Decker 2017a for analysis of a similar beat-driven cue in a non-combat film score from the *Fast and Furious* action franchise.

   11. Owen 2012, 134.

   12. Amis 2008, 53.

   13. Burgoyne 2008, 45.

   14. Prince 2009, 112.

   15. Donnelly 2014, 85.

   16. Donnelly 2014, 194.

CHAPTER TEN

   1. Basinger 1986 [2003], 5.

   2. Bronfen 2012, 150; Polan 2004, 54.

   3. Stone 1987, 19.

   4. Stone 1987, 20.

   5. Stone 1987, 20.

   6. Larson 2010, 208.

   7. Stone 1987, 56.

   8. Herr 1977 [2009], 10.

   9. Larson 2010, 11.

   10. Eric Salzman, "Samuel Barber," *HiFi/Stereo Review* 17 (October 1966): 77–89.

11. Royal S. Brown, "Recitals and Miscellany," *High Fidelity* 26 (October 1976): 132.

12. Barbara Heyman, "Samuel Barber," in *The New Grove Dictionary of Music and Musicians* (New York: Oxford University Press, 2001).

13. Nicolas Slonimsky, "Four Melodic Masterpieces of Samuel Barber," CBS Masterworks 32 11 0005, 32 11 0006, 1966.

14. Donnelly 2014, 87.

15. Kinder 2015, 8.

16. Before *Platoon*, Barber's *Adagio for Strings* had been used in three quite different films. The independent film *A Very Natural Thing* (1973) used it for an extended, lyrical lovemaking scene occurring early on in this narrative about a gay man looking for love and commitment in early-1970s Manhattan. Director David Lynch's *The Elephant Man* (1980) uses the *Adagio* for the gently presented death of John Merrick, the title character, who chooses to lie down to sleep and thereby dies of asphyxiation. The music plays before, during, and after Merrick's death—which is not shown—and continues during the film's final moments, which chart Merrick's journey into the afterlife. Director Gregory Nava inserted Barber's *Adagio* into two key scenes in *El Norte* (1983), his tale of a brother and sister fleeing political repression in Guatemala and coming to the United States in search of a better life. In Nava's first use, the Barber sneaks in behind an intense conversation between siblings Rosa and Enrique, who decide to run away together after corrupt local officials behead their father and imprison their mother. The *Adagio* expresses the resilience of their shared youth, and hope in the face of having lost everything but each other. Near the close of the film, the *Adagio* returns as Enrique sits at Rosa's hospital bed. She is dying of a form of typhus contracted from exposure to rats. The pair had crossed the US–Mexico border by crawling for miles through an unused sewer pipe. Their encounter in the pipe with hundreds of rats is sustained, horrifying, and unscored. Rosa's death much later in the narrative as a result of this episode comes as a blow: the pair had been doing well, and the viewer's hopes for their future were high. But *El Norte* offers no tales of finding a better life in America. Played behind Rosa's death, with Enrique collapsed in grief across her dead body, the meaning of Barber's *Adagio* shifts from its first use in the film. In the pair's Guatemalan home, the music suggests hope against hope; in the Los Angeles hospital, the music expresses the despair of permanent loss. The second use demonstrates the power of Barber's music as accompaniment not to narrative action or dialogue but to a tableau, stopping the action and letting the viewer contemplate a tragic moment in the story. At work on his own Latin American film *Salvador* in the early 1980s, Oliver Stone likely saw *El Norte*.

17. Delerue's unused cues can be heard on *Salvador/Platoon*, Prometheus Records PCD 136.

18. Caputo 1977, 288–89, 255, 289.

19. See Turse 2013 for historical context on the village massacre scene in *Platoon*.

20. Appy 2015, 228.

21. http://www.filmscoremonthly.com/articles/2007/15_Aug---Basil_ Poledouris_In_His_Own_Words_Part_Six.asp. Thank you to Nicholas Kmet for this reference.

22. The *Adagio* was subsequently interpolated into a variety of other genres: *Lorenzo's Oil* (1992, Hollywood melodrama), *Amèlie* (2001, French romantic comedy), *Reconstruction* (2003, Danish romantic drama), *Swimming Upstream* (2003, Australia melodrama), and *Sicko* (2007, American documentary).

23. Lang 1969.

24. For example, *Cinema Concerto: Ennio Morricone at Santa Cecilia*, Sony Classical SK 61672, 2000; and *Yo-Yo Ma Plays Ennio Morricone*, Sony Classical SK 93456, 2004.

25. Suid 2002, 612.

26. *Courage Under Fire* press kit, AMPAS.

27. Hammond 2002, 63.

28. Kovic 1976, 54.

29. Bruzzi 2013, 84.

30. The eight-minute scene with diegetic Édith Piaf records is counted here as unscored.

31. Barker 2011, 28.

32. In the book, Iggy's fate is told in a quote from the real Doc Bradley: "The Japanese had pulled him underground and tortured him. His fingernails . . . his tongue. . . . It was terrible. I've tried so hard to forget this." Bradley 2000, 344.

33. Pendleton 2009, 186.

34. *The Hurt Locker* script "third draft," AMPAS.

35. Koppl 2009a.

CHAPTER ELEVEN

1. Chion 2009, 158.

2. Winters 2010, 236.

3. Barker 2011, 120.

4. Many of these special border markings can be found in the pre-Vietnam history of the combat genre.

5. *SPR* press kit, AMPAS.

6. Kornbluth and Sunshine 1999, 84.

7. My thanks to an anonymous reader for the notion of the men as monstrous children.

8. Kassabian 2001, 3.

9. See chapter 2 for a discussion of how John Williams's end-titles music for *Born on the Fourth of July* shapes that film's final meaning.

10. David Ansen, *Newsweek*, 21 August 1989; Stuart Klawans, *The Nation*, 4 and 11 September 1989.

11. Greven 2009, 77.

12. Michael Wilmington, *LAT*, 18 August 1989.

13. Morricone's decision to develop themes associated with Oanh's ordeal contrasts with his score for another violent film set in the jungle, Roland Joffé's *The Mission* from three years earlier. Morricone's end-titles cue for *The Mission*, also with chorus and orchestra, celebrates forgiveness and mercy, developing musical themes associated during the narrative with nonviolent responses to evil—even though the advocates of nonviolence are slaughtered in the film. Morricone restores the listener to a position of hope at the close of *The Mission*, perhaps in response to the film's biblical epigraph: "The light shines in the darkness, and the darkness has not overcome it." Not so the end titles for *Casualties of War*.

14. Peebles 2011, 172.

15. Bigelow clearly likes musical hits on the titles. After starting in a restrained manner, the end title music for *Zero Dark Thirty* builds through a series of such hits toward a climactic musical arrival when the title of the film finally appears on-screen.

16. Scott Tobias, "Interview: Kathryn Bigelow," AVClub.com, http://www.avclub.com/article/kathryn-bigelow-29544. In Keough 2013, 158.

17. Two tracks by Ministry, from the same album as "Khyber Pass," are briefly played by James in his quarters during the narrative.

18. Peebles 2011, 173.

19. Pendleton 2009, 194.

20. Janet Maslin, "A Case of Honor and Lies," *NYT*, 12 July 1996.

21. *Valor: The Album, The Only Easy Day Was Yesterday*, Relativity Music Group RMG 1031–1, 2012. Urban's "For You" is the only music in *Act of Valor* on this audio paratext, which reached number 8 on *Billboard*'s Country Albums chart; number 5 on Independent Albums; and number 4 on Soundtracks.

22. Assessing the film's reception on IMDb.com, Martin Barker noted, "Fascinatingly, the film was read almost universally as pro-US. One IMDb critic of the film put it strongly: 'It's a simplistic, gung ho, neocon piece of wish fulfilment.'" Barker notes how *The Kingdom* cultivates a more subtle and ambiguous message about violence and revenge on both sides—extended here in an examination of the film's score—but concludes that for many viewers, "If American soldiers win, little else matters. Politicians can lie and deceive . . ., but if US forces do their 'action thing' and win, nothing else matters. Even knowing that you have made yet more enemies doesn't matter. 'We' won." Barker 2011, 109, 111.

23. The tune heard here occurs twice earlier in the film: during a montage crosscutting between Faris at home with his family and Fleury and the FBI team resting, and at a moment of friendship building when Fleury asks Faris what his first name is.

24. Holger Pötzsch, "Beyond Mimesis: War, History, and Memory in *Flags of Our Fathers*," in Schubart and Gjelsvik 2013, 135.

25. Eastwood also reintroduces diegetic sound at the very end of *American Sniper*: footage of the real Kyle's funeral ends with the image and sound of a SEAL pounding his pin into the lid of Kyle's coffin—repeating a trope from the

close of *Act of Valor,* only with real SEALs at a real funeral instead of real SEALs at a fake funeral.

26. Interviewed on *Fresh Air,* National Public Radio, 14 November 2007.
27. Bacevich 2013, 189.
28. *AN* press kit, AMPAS.
29. Larson 2010, 157.
30. Burgoyne 2014, 258.

Works Cited

(Newspaper and magazine articles referenced only once are listed in notes only.)

Ambrose, Stephen. 1997. *Citizen Soldiers: The U.S. Army from the Normandy Beaches to the Bulge to the Surrender of Germany, June 7, 1944, to May 7, 1945.* New York: Simon and Schuster.

Amis, Martin. 2008. *The Second Plane: September 11, Terror and Boredom.* New York: Knopf.

Appy, Christian G. 2003. *Patriots: The Vietnam War Remembered from All Sides.* New York: Viking.

———. 2015. *American Reckoning: The Vietnam War and Our National Identity.* New York: Viking.

Arkin, William M. 2015. *Unmanned: Drones, Data, and the Illusion of Perfect Warfare.* New York: Little, Brown and Company.

Aveyard, Karina, and Albert Moran, eds. 2013. *Watching Films: New Perspectives on Movie-Going, Exhibition and Reception.* Bristol, England: intellect.

Bacevich, Andrew J. 2013. *Breach of Trust: How Americans Failed with Soldiers and Their Country.* New York: Metropolitan.

Bailey, Beth. 2009. *America's Army: Making the All-Volunteer Force.* Cambridge, MA: Harvard University Press.

Barker, Martin. 2011. *A "Toxic Genre": The Iraq War Films.* London: Pluto Press.

Basinger, Jeanine. 1986 [2003]. *The World War II Combat Film: Anatomy of a Genre.* Middletown, CT: Wesleyan University Press.

Bergen, Peter L. 2012. *Manhunt: The Ten-Year Search for bin Laden from 9/11 to Abbottabad.* New York: Crown.

Bowden, Mark. 2002. "Foreword" to Ken Nolan, *Black Hawk Down: The Shooting Script.* New York: Newmarket Press.

Bradley, James, with Ron Powers. 2000. *Flags of Our Fathers.* New York: Bantam.

Brennan, Matthew. 1987. *Headhunters: Stories from the 1st Squadron, 9th Cavalry in Vietnam, 1965–1971*. New York: Presidio Press.

Bronfen, Elisabeth. 2012. *Specters of War: Hollywood's Engagement with Military Conflict*. New Brunswick, NJ: Rutgers University Press.

Bruzzi, Stella. 2013. *Men's Cinema: Masculinity and Mise-en-Scène in Hollywood*. Edinburgh, Scotland: Edinburgh University Press.

Buchanan, Sherry. 2007. *Vietnam Zippos*. Chicago: University of Chicago Press.

Burger, Jeff, ed. 2013. *Springsteen on Springsteen: Interviews, Speeches, and Encounters*. Chicago: Chicago Review Press.

Burgoyne, Robert. 2008. *The Hollywood Historical Film*. Malden, MA: Blackwell.

———. 2014. "The Violated Body: Affective Experience and Somatic Intensity in *Zero Dark Thirty*." In *The Philosophy of War Films*, edited by David LaRocca, 247–60. Lexington: University Press of Kentucky.

Buzzell, Colby. 2005. *My War: Killing Time in Iraq*. New York: G. B. Putnam's Sons.

Caputo, Philip. 1977. *A Rumor of War*. New York: Holt, Rinehart, and Winston.

Castner, Brian. 2012. *The Long Walk: A Story of War and the Life That Follows*. New York: Doubleday.

Chapman, James. 2008. *War and Film*. London: Reaktion Books.

Chion, Michel. 1999. *The Voice in Cinema*. Edited and translated by Claudia Gorbman. New York: Columbia University Press.

———. 2009. *Film: A Sound Art*. Translated by Claudia Gorbman. New York: Columbia University Press.

Clancy, Tom. 2012. "Foreword" to Dick Crouch and George Galdorisi, *Tom Clancy Presents Act of Valor*. New York: Berkley.

Cockburn, Andrew. 2015. *Kill Chain: The Rise of the High-Tech Assassins*. New York: Henry Holt.

Cooper, Marc. 1988. "*Playboy* Interview: Oliver Stone." In *Oliver Stone: Interviews*, edited by Charles L. P. Silet, 59–90. Jackson: University Press of Mississippi.

Cowie, Peter. 2001. *The Apocalypse Now Book*. New York: Da Capo Press.

Crouch, Dick, and George Galdorisi. 2012. *Tom Clancy Presents Act of Valor*. New York: Berkley.

Cusick, Suzanne. 2008. "'You Are in a Place That Is Out of the World . . .': Music in the Detention Camps of the 'Global War on Terror.'" *Journal of the Society for American Music* 2 (1): 1–26.

Daughtry, J. Martin. 2015. *Listening to War: Sound, Music, Trauma, and Survival in Wartime Iraq*. New York: Oxford University Press.

Decker, Todd. 2016. "A Waltz with and for the Greatest Generation: Music in *Band of Brothers* (2001)." In *American Militarism on the Small Screen*, edited by Stacy Takacs and Anna Froula, 93–108. New York: Routledge.

———. 2017a. "Racing in the Beat: Music in the *Fast and Furious* Franchise." In *Contemporary Musical Film*, edited by K. J. Donnelly and Beth Carroll. Edinburgh, Scotland: Edinburgh University Press.

———. 2017b. "The Filmmaker as DJ: Martin Scorsese's Compiled Score for *Casino* (1995)." *Journal of Musicology* 34 (2).

Dienstfrey, Eric. 2016. "The Myth of the Speakers: A Critical Reexamination of Dolby History." *Film History* 28 (1): 167–93.

Donnelly, K.J. 2014. *Occult Aesthetics: Synchronization in Sound Film.* New York: Oxford University Press.

Dupuy, Colonel R. Ernest, and Lieutenant Colonel Herbert L. Bregstein. 1946. *Soldiers' Album.* Boston: Houghton Mifflin.

Engel, Richard. 2008. *War Journal: My Five Years in Iraq.* New York: Simon and Schuster.

Filkins, Dexter. 2008. *The Forever War.* New York: Knopf.

Finkel, David. 2009. *The Good Soldiers.* New York: Sarah Crichton Books.

———. 2013. *Thank You for Your Service.* New York: Sarah Crichton Books.

Ford, Daniel. 1967. *Incident at Muc Wa.* New York: William Heinemann.

Friedman, SGM Herbert A. (Ret.) n.d. "The Use of Music in Psychological Operations." http://www.psywarrior.com/MusicUsePSYOP.html.

Gengaro, Christine Lee. 2012. *Listening to Stanley Kubrick: The Music in His Films.* New York: Rowman and Littlefield.

Gilman, Lisa. 2016. *My War, My Music: The Listening Habits of U.S. Troops in Iraq and Afghanistan.* Middletown, CT: Wesleyan University Press.

Gray, J. Glenn. 1959. *The Warriors: Reflections on Men in Battle.* New York: Harcourt, Brace.

Greengrass, Paul. 2006. *United 93: The Shooting Script.* New York: Newmarket Press.

Greven, David. 2009. *Manhood in Hollywood from Bush to Bush.* Austin: University of Texas Press.

Hagedorn, Ann. 2014. *The Invisible Soldiers: How America Outsourced Our Security.* New York: Simon and Schuster.

Hammond, Michael. 2002. "Some Smothering Dreams: The Combat Film in Contemporary Hollywood." In *Genre and Contemporary Hollywood.* Edited by Stephen Neale. London: BFI Publishing.

Hasford, Gustav. 1979. *The Short-Timers.* New York: Harper and Row.

Hellmann, John. 1986. *American Myth and the Legacy of Vietnam.* New York: Columbia University Press.

Hemphill, Robert. 1998. *Platoon: Bravo Company.* Fredericksburg, VA: Sergeant Kirkland's Press.

Herr, Michael. 1977 [2009]. *Dispatches.* New York: Alfred A. Knopf. Reprinted 2009 with an introduction by Robert Stone in Everyman's Library.

Hillstrom, Kevin, and Laurie Collier Hillstrom. 1998. *The Vietnam Experience: A Concise Encyclopedia of American Literature, Songs, and Films.* Westport, CN: Greenwood Press.

Hinson, Hal. 1987. "The Rugged *Hill*." *Washington Post*, 28 August.

Huebner, Andrew J. 2008. *The Warrior Image: Soldiers in American Culture from the Second World War to the Vietnam Era.* Chapel Hill: University of North Carolina Press.

Inglis, Ian. 2005. "Music, Masculinity, and Membership." In *Pop Fiction: The Song in Cinema*, edited by Steve Lannin and Matthew Caley, 63–70. Bristol, England: intellect.

Jackson, Blair. 1979. "*Apocalypse Now*: The Music, The Making of a Remarkable Film Score." *BAM Magazine: The California Music Magazine*, 5 October, 24–30.

Jorgenson, Kregg P. J. 1991. *Acceptable Loss: An Infantry Soldier's Perspective*. New York: Ivy Books.

Junger, Sebastian. 2010. *War*. New York: Twelve.

Kassabian, Anahid. 2001. *Hearing Film: Tracking Identifications in Contemporary Hollywood Film Music*. New York: Routledge.

Kennedy, David M., ed. 2013. *The Modern American Military*. New York: Oxford University Press.

Kennedy, Kelly. 2010. *They Fought for Each Other: The Triumph and Tragedy of the Hardest Hit Unit in Iraq*. New York: St. Martin's Press.

Keough, Peter, ed. 2013. *Kathryn Bigelow: Interviews*. Jackson: University Press of Mississippi.

Kieran, David. 2014. *Forever Vietnam: How a Divisive War Changed American Public Memory*. Amherst: University of Massachusetts Press.

Kinder, John M. 2015. *Paying with Their Bodies: American War and the Problem of the Disabled Veteran*. Chicago: University of Chicago Press.

King, Geoff. 2000. *Spectacular Narratives: Hollywood in the Age of the Blockbuster*. New York: I. B. Tauris.

Kinney, Katherine. 2000. *Friendly Fire: American Images of the Vietnam War*. New York: Oxford University Press.

Koppl, Rudy. 2009a. "The Hurt Locker: Blurring the Lines between Sound and Score." http://www.musicfromthemovies.com/index2.php?option=com_content&view=article&id=63.

———. 2009b. "The Hurt Locker: Mainlining the War at Death's Door." http://www.musicfromthemovies.com/index2.php?option=com_content&view=article&id=64.

Kornbluth, Josh, and Linda Sunshine. 1999. "*Now You Know*": *Reactions after Seeing Saving Private Ryan, Compiled by America Online and DreamWorks*. New York: Newmarket Press.

Kovic, Ron. 1976. *Born on the Fourth of July*. New York: McGraw Hill.

Kubrick, Stanley, Michael Herr, and Gustav Hasford. 1987. *Full Metal Jacket: The Screenplay*. New York: Random House.

Lang, Daniel. 1969. "Casualties of War." *New Yorker*, 18 October, 61–146.

Langford, Barry. 2010. *Post-Classical Hollywood: Film Industry, Style and Ideology since 1945*. Edinburgh, Scotland: Edinburgh University Press.

Larson, Thomas. 2010. *The Saddest Music Ever Written: The Story of Samuel Barber's Adagio for Strings*. New York: Pegasus.

Leckie, Robert. 1957. *Helmet for My Pillow: From Parris Island to the Pacific*. New York: Random House.

Lichtenfeld, Eric. 2007. *Action Speaks Louder: Violence, Spectacle, and the American Action Movie*. Rev. and expanded ed. Middletown, CT: Wesleyan University Press.

Luttrell, Marcus, with Patrick Robinson. 2007. *Lone Survivor: The Eyewitness Account of Operation Redwing and the Lost Heroes of Seal Team 10*. New York: Little, Brown.

Marlantes, Karl. 2010. *Matterhorn: A Novel of the Vietnam War*. Berkeley: El León Literary Arts.

McCain, Cilla. 2010. *Murder in Baker Company: How Four American Soldiers Killed One of Their Own*. Chicago: Chicago Review Press.

McChrystal, General Stanley. 2013. *My Share of the Task: A Memoir*. New York: Portfolio.

McDonald, Paul, and Janet Wasko, eds. 2008. *The Contemporary Hollywood Film Industry*. Malden, MA: Blackwell.

McManus, John C. 2010. *Grunts: Inside the American Infantry Combat Experience, World War II Through Iraq*. New York: NAL Caliber.

McNeill, William H. 1995. *Keeping Together in Time: Dance and Drill in Human History*. Cambridge, MA: Harvard University Press.

McQuiston, Kate. 2013. *We'll Meet Again: Musical Design in the Films of Stanley Kubrick*. New York: Oxford University Press.

Modine, Matthew. 2005. *Full Metal Jacket Diary*. New York: Rugged Land.

Murch, Walter. 2001. *In the Blink of an Eye: A Perspective on Film Editing*. 2nd edition. Los Angeles: Silman-James Press.

Nolan, Ken. 2002. *Black Hawk Down: The Shooting Script*. New York: Newmarket Press.

O'Brien, Wesley J. 2012. *Music in American Combat Films: A Critical Study*. Jefferson, NC: McFarland.

Owen, Mark [Matt Bissonnette], with Kevin Maurer. 2012. *No Easy Day: The Autobiography of a Navy SEAL*. New York: Dutton.

Peebles, Stacey. 2011. *Welcome to the Suck: Narrating the American Soldier's Experience in Iraq*. Ithaca, NY: Cornell University Press.

Pelfrey, William. 1987. *Hamburger Hill: A Novel Based on the Screenplay Written by Jim Carabatsos*. New York: Avon Books.

Pendleton, David. 2009. "Discussion at Harvard Film Archive." In *Kathryn Bigelow: Interviews*. Edited by Peter Keough. Jackson: University Press of Mississippi, 2013.

Pfeil, Fred. 2004. "Terrence Malick's War Film Sutra: Meditating on *The Thin Red Line*." In *New Hollywood Violence*. Edited by Steven Jay Schneider. Manchester, England: Manchester University Press.

Pieslak, Jonathan. 2009. *Sound Targets: American Soldiers and Music in the Iraq War*. Bloomington: Indiana University Press.

Polan, Dana. 2004. "Auterism and War-teurism: Terrence Malick's War Movie" (1998). In Robert Eberwein, *The War Film*. New Brunswick, NJ: Rutgers University Press. Originally published in *Meteor* 14. Vienna: PVS Verleger.

Powers, Kevin. 2012. *The Yellow Birds*. New York: Little, Brown.

Prince, Stephen. 2009. *Firestorm: American Film in the Age of Terrorism*. New York: Columbia University Press.

Psujek, Jennifer. 2016. "The Composite Score: Indiewood Film Music at the Turn of the Twenty-First Century." PhD diss., Washington University in St. Louis.

Purse, Lisa. 2011. *Contemporary Action Cinema*. Edinburgh, Scotland: Edinburgh University Press.

———. 2013. *Digital Imaging in Popular Cinema*. Edinburgh, Scotland: Edinburgh University Press.

Robinson, Roxana. 2013. *Sparta: A Novel*. New York: Sarah Crichton Books.

Savage, Kirk. 2009 [2011]. *Monument Wars: Washington, D.C., the National Mall, and the Transformation of the Memorial Landscape*. Berkeley: University of California Press.

Schubart, Rikke, and Anne Gjelsvik, eds. 2013. *Eastwood's Iwo Jima: Critical Engagement with "Flags of Our Fathers" and "Letters from Iwo Jima."* New York: Wallflower Press.

Scruggs, Jan C., and Joel L. Swerdlow. 1985. *To Heal a Nation: The Vietnam Veterans Memorial*. New York: Harper Collins.

Stone, Oliver. 1987. *Oliver Stone's "Platoon" and "Salvador."* New York: Vintage.

Stubbs, Jonathan. 2013. *Historical Film: A Critical Introduction*. New York: Bloomsbury.

Suid, Lawrence. 2002. *Guts and Glory: The Making of the American Military Image in Film*. Rev. and expanded ed. Lexington: University Press of Kentucky.

Swofford, Anthony. 2003. *Jarhead: A Marine's Chronicle of the Gulf War and Other Battles*. New York: Scribner.

Théberge, Paul. 2008. "Almost Silent: The Interplay of Sound and Silence in Contemporary Cinema and Television." In *Lowering the Boom: Critical Studies in Film Sound*. Edited by Jay Beck and Tony Grajeda. Urbana: University of Illinois Press.

Tregaskis, Richard. 1943 [2000]. *Guadalcanal Diary*. New York: Modern Library.

Turner, Brian. 2005. *Here, Bullet*. Farmington, ME: Alice James Books.

———. 2010. *Phantom Noise*. Farmington, ME: Alice James Books.

Turse, Nick. 2013. *Kill Anything That Moves: The Real American War in Vietnam*. New York: Metropolitan Books.

"Walter Murch." 1996. In *Projections 6*, edited by John Boorman and Walter Donohue, 149–62. London: Faber and Faber.

West, Bing. 2005. *No True Glory: A Frontline Account of the Battle for Fallujah*. New York: Bantam.

Winters, Ben. 2010. "The Non-Diegetic Fallacy: Film, Music, and Narrative Space." *Music and Letters* 91 (2): 224–44.

Wong, Leonard. 2005. "Leave No Man Behind: Recovering America's Fallen Warriors." *Armed Forces and Society* 31 (4): 599–622.

Wright, Evan. 2004. *Generation Kill: Devil Dogs, Iceman, Captain America, and the New Face of American War*. New York: G. B. Putnam's Sons.

Zaffiri, Samuel. 1988. *Hamburger Hill: May 11–20, 1969*. New York: Pocket Books.

Index

Note: Prestige combat film titles are given *in bold*. Boldface **page numbers** denote detailed description or analysis of given film and topic.